Aerodynamic Characteristics extract Analysis of a Single Tethered Aerostat

William Cox

Contents

FOREWORD

Wind data is an inevitable part of the site selection, design, development and maintenance of wind energy harvesting systems. The conventional way of collecting the wind data, such as the mast-based method, has some limitations in its altitude of operation, altitude variability, reusability, and so on. There are several drawbacks to the conventional methods, such as altitude constraints, installation difficulties, relocation limitations and so on. This research proposes a single-tethered aerostat-based wind data collection system as an alternative method. The advantages of the proposed system over the conventional methods are: it can operate at different altitudes, its altitude of operation can be easily varied, and it can be used again for a different application/location.

The proposed system consists of a single tether connecting the streamline-shaped aerostat carrying the payload to the ground station. The aerostat is a lighter-than-air (LTA) class of aerial vehicle working on the principle of buoyancy of the LTA gases filled in it. Tether constraints the aerostat from moving away and can be used as a communication and/or power channel for the payload. The payload varies depending on the application.

The focus of this research is to conduct a feasibility study on the proposed aerostat-based system. A computational fluid dynamics (CFD) based analysis methodology is adopted for the research. The aerodynamic characteristics of the aerostat under unsteady wind conditions are studied using a 3D model of the aerostat and CFD tools. Wind gust models are used for simulating the unsteady conditions of the wind. The amplitude and time period of the wind gusts is selected based on the wind data provided by the National Institute of Wind Energy (NIWE) at an altitude of 100 m above sea level. The effect of wind gust amplitude, gust time period, angle of attack (AoA), gust profile and the shape of the aerostat on the aerodynamic characteristics of the aerostat is investigated. A scaling methodology is proposed for the aerostat responses, which helps in an easy analysis and provides a way to predict the response of the aerostat to new wind gust conditions.

The stability derivatives and added mass terms are significant in the dynamics of LTA vehicles and underwater vehicles compared to the heavier-than-air vehicles. The current research investigates the stability derivatives of the aerostat under steady as well as unsteady wind and motion conditions. Continuous periodic wind gust models are used for the analysis. The unsteady condition of the aerostat is realised by the combination of unsteady periodic wind gusts and unsteady periodic motion of the aerostat. A CFD-based method for estimating the stability derivatives of the aerostat is investigated. A set of validation studies are presented to validate the methodology used, the motion of the model and the stability derivatives estimated using CFD analysis. The effect of AoA, wind gust amplitude of oscillation, frequency of oscillation, the phase shift between the wind and model oscillation, and the aerostat model oscillations on the stability derivatives are investigated in the current research.

Being a complex dynamic system involving mechanical and aerodynamic components, the tethered aerostat has to be analysed for its operation, performance and stability. The best way to do this is to represent the dynamics in a mathematical form, a mathematical model. A decoupled longitudinal model of the single-tethered aerostat is investigated in the current research. The model parameters of the aerostat (added mass terms and stability derivatives) are estimated using the CFD-based analysis. A two-dimensional tether model is presented and validated with experimental data available in the literature. The tether profile (two-dimensional coordinates of the entire tether length) is analysed for different wind velocities and tether configurations. The tether model is coupled with the aerostat equations of motion and solved using the ODE45 solver available with MATLAB 2020b. The effect of wind velocity on the aerostat response is also investigated.

Chapter 1

Introduction

1.1 Background study

Wind energy is one of the sustainable and greener renewable energy sources. In the last few decades, it has earned great interest among scientists, researchers, and industrialists all over the world. An illustration of the development of windmills over the last three decades is given in Figure 1.1[1]. As part of achieving India's goal of 60% non-fossil fuel based installed electric capacity by 2030, the country's wind energy sector is showing tremendous growth [2]. India's NIWE has estimated the wind power potential of the country at an altitude of 100 m above ground level at 302 GW [3]. This shows the importance of designing, developing, maintaining, and expanding the wind energy conversion systems in the Indian power system.

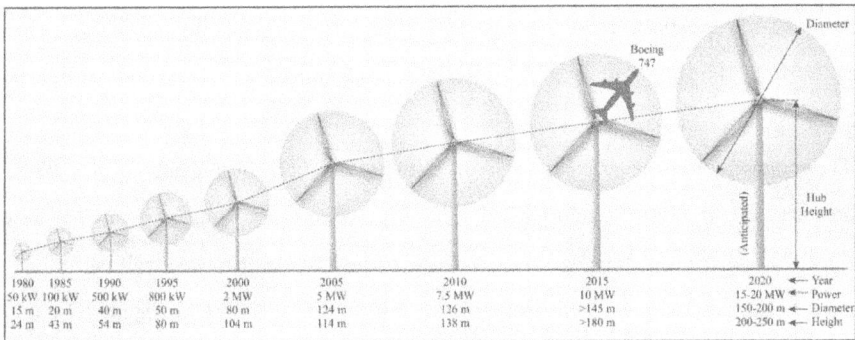

Figure 1.1: Evolution in the windmill dimensions [1].

Wind data plays a supervisory role in wind energy conversion systems. For the successful operation of wind farms, wind data for different time horizons is needed. The local wind climate,

1

turbulence levels, and expected return levels of wind gusts are needed in the planning phase. During operation, both short-range and long-range wind data are needed. Long-range (1-2 days) data will be used for operations planning and unit allotment [4]. Short-range (minutes) data will be used for detecting patterns and direction changes that may cause structural damage to the turbines [5]. The process of primary site selection is the foremost step in wind farm planning. The selection factors for a reliable site include the abundance of wind, stable direction of flow, concentrated distribution of wind, less destructive gusts, marine and aerial traffic, geography, and access to the load centre. Being a site-specific source of energy, the potential sites for wind farms are selected after an extensive wind resource assessment. It includes the speed and direction of the wind, variation of the wind with altitude, presence of turbulence, and so on, which have to be measured for a period of at least five years [6]. This data will be used as the basis for the design and development of wind turbines, the selection of the tower height, the alignment of the wind farm, and the expansion and upgrade of the existing wind farms.

Wind data can be extracted using different sensors such as cup anemometers, light detection and ranging (LiDAR), sound detection and ranging (SoDAR), synthetic aperture radar (SAR), and aerial vehicles, as shown in Figure 1.2 (a). Being the traditional and widely used wind sensor, mechanical anemometers give accurate results with minimum complexity and cost. LiDARs can be set up at lower altitudes than anemometers. Even though the measuring accuracy matches that of conventional cup anemometers, LiDARs are not used for measuring short-duration wind gusts. This is because of the time taken between the hitting and reflecting of signals from the target, and the measurement is taken over a volume of air rather than a particular point.

The common practices for collecting wind data are in situ measurements using a mast, airborne platforms, and satellite information. Conventional sensor holding masts are the primary consideration for wind profile generation for wind farms. Different sensors will be equipped at different heights of the mast to collect data for a location, as shown in Figure 1.2 (b). The problems with the masts are height constraints, location constraints, limited altitude variation, less chance of relocation, and difficulties in inspecting and maintaining the overhead sensors.

Airborne solutions have emerged to overcome the difficulties with masts. Severe weather

2

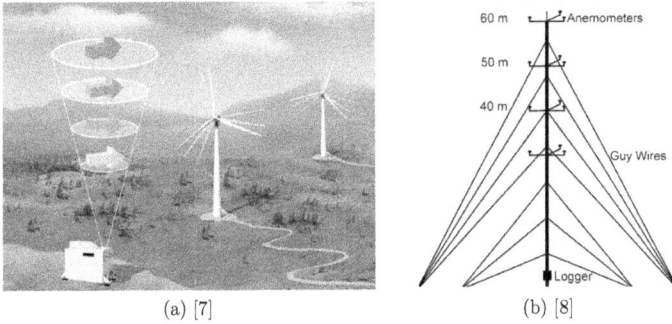

Figure 1.2: Wind data collection methods: (a) Lidar based measurement and (b) mast based method.

conditions, such as cyclones, are monitored with the help of the Doppler radar system using the unmanned aerial system. The drawbacks of airborne wind measurement systems using UAVs and other platforms are excessive power consumption, the inability to fly for a long period of time, and imposed aviation regulations.

Satellite-based wind data are usually used for the preliminary search for potential sites because the data are valid at 10 m above sea level. The efficiency of the satellite data can be improved by blending multiple satellite data sets with the help of proper interpolation algorithms.

The conventional method of wind measurement for wind farms is the mast-based method [9]. A long mast (sometimes a tower structure, depending on the terrain and wind conditions) will be installed in the identified locations with supporting tie wires. Wind measuring instruments such as a cup anemometer, LiDAR, or radar will be equipped on the mast at different heights. The conventional mast-based method for wind measurement needs a permanent structure of mast and tie wire assembly. It needs a site permit from the authorities, which may take months to obtain [4]. It is practically impossible to relocate and use the setup at a different location. Another drawback is the limited height. It requires a strong foundation and support wires for increased heights. It is very difficult to perform maintenance work at that height. There is also limited scope for varying the altitude of measurement. A rough calculation for buying and installing a 100 m met mast may cost 5.8 to 9.5 million rupees [10].

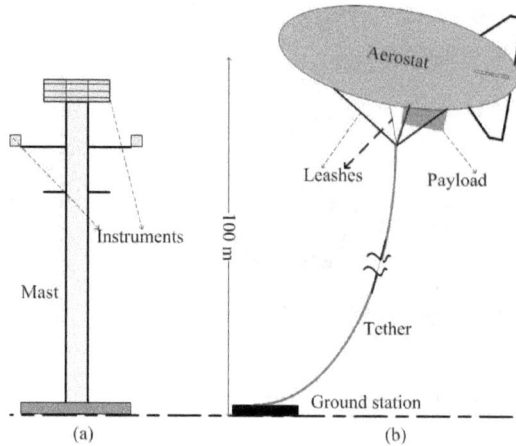

Figure 1.3: (a) Conventional and (b) proposed wind measurement system.

1.2   Research proposal

A low-altitude tethered aerostat-based wind measurement system is proposed in this research. In the proposed system, instead of installing a mast for setting up the weather station, an aerostat filled with LTA gas, such as helium or hydrogen, will be kept afloat at the identified location. The aerostat will be tethered to the ground. This tether can also act as a communication channel between the ground station and the overhead equipment. The weather station can be hung from the aerostat using leashes. The height of the aerostat from the ground is easily adjustable using the tether. The proposed tethered aerostat system can be represented as shown in Figure 1.3.

Tethered aerostats can be found in various applications, starting from recreation shows to some strategic applications, such as military surveillance and imaging. The noiseless and vibration-free operation, less energy requirement, ability to hover for a long duration, and ease of installation and dismantling compared to other aerial platforms and structures make aerostats desirable for certain applications that demand slow manoeuvring and station keeping. Some of the applications of aerostats are shown in Figure 1.4. They include the Indian Air Force's Akashdeep radar system (Figure 1.4 (a)), aerial communication and internet connectivity (Figure 1.4 (b)), Commonwealth Games 2010 (Figure 1.4 (c)), atmospheric parameter monitoring (Figure 1.4 (d)), and an aerial surveillance system (Figure 1.4 (e)).

The major benefits of the proposed system over conventional methods are:

4

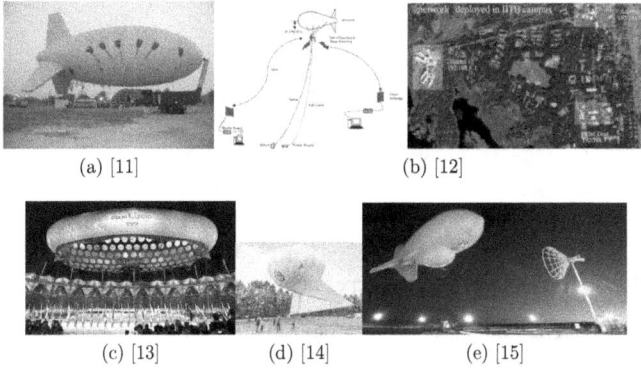

Figure 1.4: Major reported applications of aerostats: (a) Radar, (b) communication, (c) recreation, (d) atmospheric measurements, and (e) aerial surveillance.

- Raising time - Setting up the measurement system takes comparatively less time.

- Portability - Easy to relocate the entire system to other locations when dismantled.

- Altitude range - Altitude of operation can be varied over a wide range.

- Active period - Suitable for a long period of operation in a feasible weather.

- Land terrain independence - Easily implemented in soft soil or water bodies.

- Access - Easily unmounted for maintenance purpose.

The current research focuses on the study of the feasibility of the proposal using numerical simulation techniques, the investigation of the mathematical model parameters of the proposed system, and the analysis of the system with the help of a complete mathematical model.

1.3 Organisation of the thesis

The thesis has seven chapters, and the organisation is as follows:

Chapter 1 presents the necessary background information to explain the need for the research proposal. It includes the relevance of the current research, state-of-the-art techniques, their drawbacks, and the proposed system.

Chapter 2 discusses the relevant literature on the research topic to understand the theory, state-of-the-art methodologies, research gaps identified, and the research objectives of the current research.

Chapter 3 presents the research methodology adopted for the current research by the author based on the literature review. A brief explanation of the sequential steps undertaken in the current research is presented. These include the aerostat geometry, wind gust models, the simulation environment setup for the current research, and the stability derivative extraction methodology.

Chapter 4 discusses the aerodynamic performance analysis of the aerostat under unsteady wind conditions. Different wind gust models from the literature, four AoA, different amplitudes and time periods for the gust models, and some geometrical attributes of the aerostat were used for this analysis.

Chapter 5 presents the stability derivative extraction study of the aerostat using CFD. It includes the study of the aerostat under steady and unsteady wind conditions.

Chapter 6 presents the mathematical modelling of the tethered aerostat system. A 2-DoF model of the tether and harness and the decoupled longitudinal dynamic model of the aerostat are presented.

Chapter 7 concludes the thesis with major findings and the future scope of the current research.

Chapter 2

Literature Review

The aim of this research is to investigate the feasibility of the proposal of a single-tethered aerostat for the measurement of wind at an altitude of around 100 m above ground level. The current research started with the search for an alternate method for wind measurement at wind farm sites. A single-tethered aerostat-based system was found to be a considerable choice for the investigation. The current research focuses on the analysis and modelling of the above-mentioned aerostat system. In this chapter, a literature review of the areas considered for the current research is presented. The literature considered is classified into three categories: study of the aerostat, analysis of the effect of wind on the aerostat, and mathematical modelling of the tethered aerostat. Based on the observations from the literature review, some research gaps are identified, and the research objectives of the current research are presented in the concluding section of this chapter.

2.1 Study of the aerostat

This section presents the fundamental ideas regarding a tethered aerostat, including the history of the aerostat, the principle of operation, the classification of the aerostat, and the applications of the aerostat.

2.1.1 Fulfilling the dream to fly: The history of aerostats

Man has possessed the dream of flying like a bird since time immemorial. Even before understanding the aerodynamic principles behind a winged flight, the idea of buoyancy was

(a) [16] (b) [16]

Figure 2.1: A spherical hot air balloon with manned gondola.

identified by the common man in the historical era. In China, they have used sky lanterns as part of the tradition as well as for signalling [17]. The first manned flight of a hot air balloon was in 1783 by Jean-Francois Pilatre de Rozier [18]. There was a time when scientists competed to reach new heights using their own designed hot air balloons, such as the one seen in the 2019 movie 'The Aeronauts', as shown in Figure 2.1 [16]. With the advancement of science and technology, the source of buoyancy was changed to LTA gases such as hydrogen and helium rather than hot air [19]. The developments in the materials used for the envelope and the aerodynamic design of the envelope shape caused the bloom in the hot air balloon systems and aerostats.

Aerostat is the generic name given to a class of aerial vehicles or platforms that work on the principle of aerostatic lift. Hot air balloons, tethered aerostats, airships, and helikites are the categories of aerostats. The First World War made the major powers of the world adopt new strategies to fight, such as the use of aerial vehicles. The low operating cost of the aerostats and the immature aircraft technology of that time caused the bloom of the application of aerostats in the war. There started the golden age of aerostats, which gradually attained global acceptance for various applications.

Airships were used for strategic as well as transportation purposes until the first half of the 20th century. There were several manned flights that happened in the major cities of London and the United States. The poor ground handling facilities and the flammable nature of the hydrogen gas led to a series of catastrophic accidents with the airships. The Hindenburg disaster in 1937 was one of the unforgettable accidents in which 35 human lives were lost. By the middle of the 20th century, a decline in interest and technological developments in aerostats

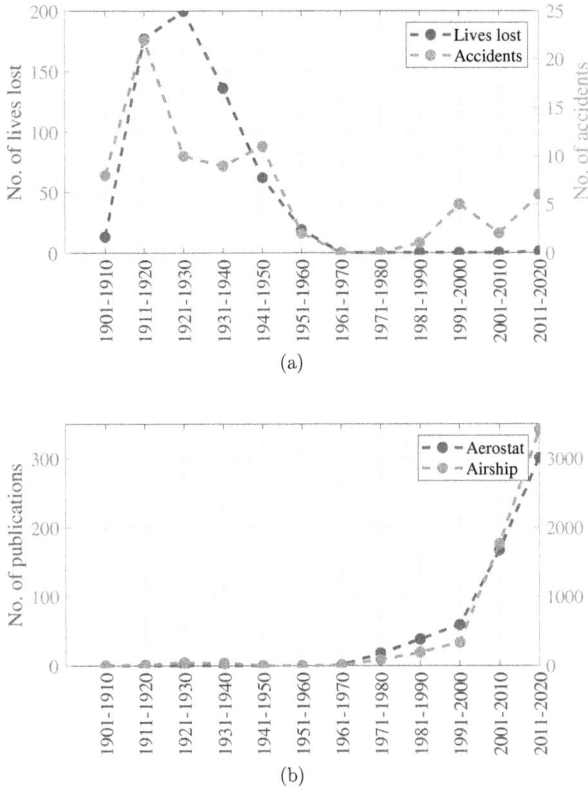

Figure 2.2: Number of aerostat (a) accidents and (b) aerostat related publications.

had started. The primary reasons were the higher chance of accidents and the advancements in aircraft technology. Figure 2.2 (a)[1] shows the number of accidents that happened with the aerostat systems.

The search for a platform with less noise, less speed, and less vibration for strategic applications such as military surveillance again led researchers and industrialists towards aerostat technology. Towards the end of the 20th century, aerostat technology regained its lost potential among academicians and researchers, as is evident by the number of publications related to aerostat technology, as shown in Figure 2.2 (b)[2].

---

[1]Data source: Wikipedia
[2]Data source: Scopus

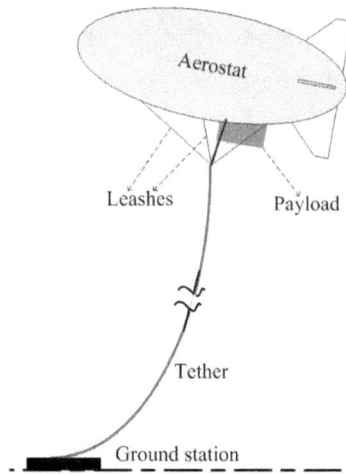

Figure 2.3: Schematic diagram of a single-tethered aerostat.

2.1.2 Principle of operation

Aerostats are a class of aerial vehicles/platforms that work on the principle of aerostatic lift provided by the buoyancy of the LTA gas. Hot air, hydrogen, and helium are the most widely used LTA gases. Due to the regulations imposed on hydrogen because of its flammable property, it is seldom used for manned flights. In general, helium is commonly used in aerostats for various applications.

As the current research involves the analysis of a tethered aerostat, this section presents the principle of operation of a tethered aerostat. A pictorial representation of a tethered aerostat is shown in Figure 2.3. The components of the tethered aerostat may vary based on the operating conditions and applications. Nonetheless, a basic tethered aerostat has a huge gas envelope called a hull, control surfaces in the form of tail fins, tethers, and a payload. A brief discussion of each of these is given in the coming sections.

As mentioned earlier, the static lift for the aerostat is provided by the large volume of LTA gas. The envelope holding the gas is known as the hull of the aerostat. The payload carrying capacity of the aerostat is highly dependent on the overall weight of the aerostat system. The shape of the hull plays an important role in the aerodynamic characteristics of the aerostat as well. The common shapes reported for the aerostat hull are spherical and streamline shapes. Being the geometry with the least surface area for a given volume, the

Table 2.1: A comparison between spherical and streamline aerostats.

	Sphere	Streamlined
Area of cross section	Constant	Variable
Front area	More for same volume	Less for same volume
Drag	Large drag coefficient. High lift efficiency	Smaller drag coefficient. Sensitive to cross winds
Ground handling	Comparatively simple	Complex
Weather vaning	No need (since symmetric)	Need to be considered (since asymmetric geometry)
Response to wind	More position error due to wind gusts.	Less position error due to wind gusts.
Availability	Not easily available	Easily available commercially
Cost	Comparatively cheaper	Added cost due to control surfaces
Design	Not a common design uncertainty in design, fabrication and operation	Well established design and fabrication methodology

spherical shape can guarantee the minimum weight for the hull. The streamline shape with better drag characteristics compared to the spherical shape has the added advantage of being able to align with the wind. A comparison between these two shapes is given in Table 2.1. There are other shapes reported in the literature for the aerostat hull, such as spherical [20, 21] (Figure 2.4 (a)), kite [22] (Figure 2.4 (b)), annular shroud [23] (Figure 2.4 (c)), ellipsoid [24, 25, 26] (Figure 2.4 (d)), oblate spheroid [27, 28] (Figure 2.4 (e)), winged [29] (Figure 2.4 (f)), and streamline [21, 30, 31, 32, 33] (Figure 2.4 (g)). They are application-oriented shapes; for example, Figure 2.4 (c) is for wind energy harvesting.

Apart from the aerodynamic shape of the hull, the aerodynamic properties of the aerostat mainly depend on the tail fins. They can act as active or passive control surfaces for the aerostat. The tails can be built of the same material as the aerostat and inflated. There are rigid fins that are made of different materials so that they can be used as active control surfaces. The orientation and number of the tail fins differ based on the degree of control and application. Three- and four-finned aerostats are commonly available in the literature. They can be oriented as 'X', '+', 'Y', or inverted 'Y'. There are works reported that the enlarged wing can help the aerostat to sustain in very high winds [29].

A tether is the mechanical connection between the aerostat and the ground station in the form of a cable or rope. Sometimes, the tether acts as the power and communication channel

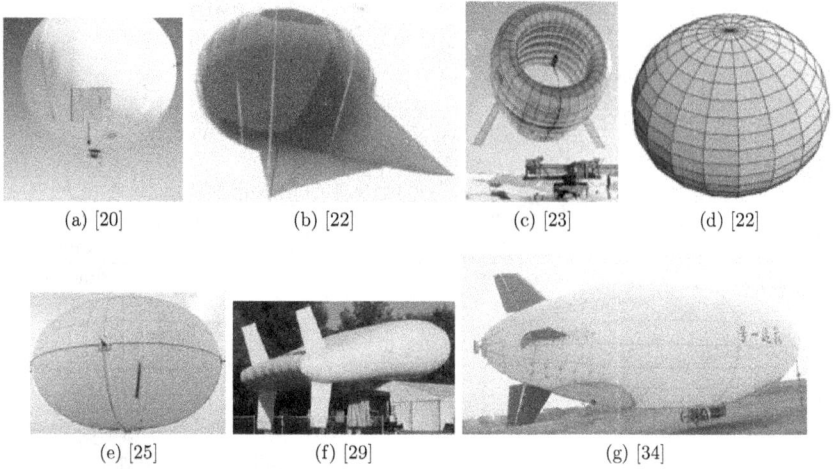

Figure 2.4: Aerostat shapes reported in literature: (a) spherical, (b) kite shaped, (c) annular shroud, (d) ellipsoid, (e) oblate spheroid, (f) winged, and (g) streamline.

between the aerostat and the ground station. The number of tethers can be single or multiple. As the number of tethers increases, the station-keeping capability of the aerostat improves at the expense of the space required for the entire system. Single tethered aerostats are commonly used because of their low space requirement and ease of operation. Controlling the length of the tether is the most common control action involved in tethered aerostat-based systems. Mechanical winches are used to release and retract the tether based on the change in tension or the change in altitude.

The payload of the aerostat is application-dependent. For an aerial imaging application, the payload can be cameras, data storage elements, power sources, communication devices, and the platform to hold the equipment together. Based on the accuracy required for the application and the potential of the payload, the performance of the aerostat has to be controlled. A highly sensitive payload may demand precise positioning of the aerostat, which will make the control complex. The resolution and sensitivity of the payload should be selected based on the characteristics of the aerostat.

Aerostats can be classified based on their shapes, source of power, operating conditions, applications, structure, and so on. In this research, an elementary classification, as shown in Figure 2.5, is presented. The current research is focused on the tethered aerostat without any propulsion sources equipped.

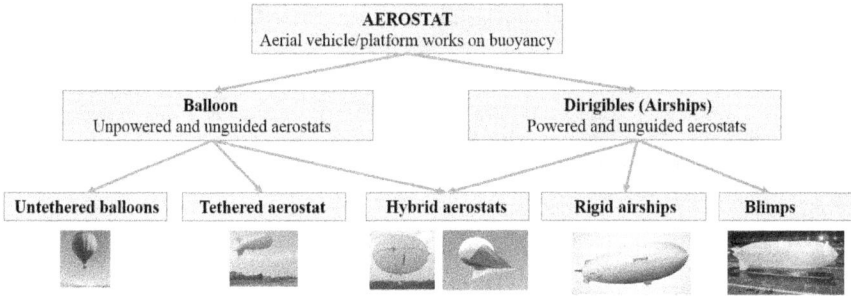

Figure 2.5: Classification of aerostats.

2.2   Analysis of the effect of wind on aerostat

As the application is mostly in open atmospheric conditions, the aerostat will be in continuous interaction with atmospheric parameters such as the wind. The buoyancy effect and the large surface area/volume of the aerostat make it highly susceptible to variations in the wind. In this section, a review of the studies related to the wind analysis of the aerostat is presented. It includes steady as well as unsteady conditions. As the current research deals with a low-altitude application, the literature considered is mostly from low-altitude cases. The literature on the analysis of the effect of wind on the aerostat can be found based on flight tests, experimental studies using wind tunnels or tow tanks, and numerical studies.

The flight test-based studies reported in the literature mostly addressed the stability and tension of the tethers. For instance, a steady flight test conducted using the 420K streamline aerostat has investigated the tether tension characteristics, stability, and validation of a theoretical study [35]. Jones and Krausman conducted a verification study of a non-linear model of a tethered aerostat using the TCOM aerostat [36]. They have reported significant sources of error in measurements while performing the flight test of a tethered aerostat. Lambert presented a comparative study on three different types of aerostats and reported that the streamline aerostat is more aerodynamically stable than its spherical counterpart [37]. A flight test of a spherical aerostat was reported by Nahon to investigate its dynamic characteristics and to develop a mathematical model of the tether and aerostat [38]. A multi-tethered aerostat-based position control system was reported, which was based on a flight test, and the disturbance rejection capability of various controllers was investigated experimentally [39]. There are some other works reported by Nahon based on the tethered aerostat-based experimental studies,

13

which include controller design and simulation model validation [40, 41, 42]. Pant has reported a flight test of a spherical and streamline aerostat for the feasibility study of a wireless internet application [12]. They have also reported a payload recovery testing study based on a flight test of an aerostat [24]. There are some studies on the tow tank and wind tunnel-based experimental studies of aerostats as well [43, 44, 45, 46, 47]. An unconventional aerostat prototype was tested in a water channel for obtaining the model parameters by Deodhar by considering different heading angles and moment centres [28].

The advanced developments in the computational power of computers have revolutionised numerical computation applications in engineering fields. CFD is one of the most well-developed and widely applied design and analysis platforms for investigating fluid-structure interaction. The physical complexities, high cost, practical limitations of the tow tanks and wind tunnels, and difficulties in obtaining all the required characteristic responses during experiments have increased the application of CFD among academicians, industrialists, and researchers [48]. There are some works that report the application of CFD for investigating the aerodynamic interaction between wind and the aerostat. Most of them dealt with the calculation of aerodynamic force and moment coefficients for a range of AoA [47, 49, 50]. Those works have considered both negative and positive AoA. For instance, the effect of turbulence was studied on a scaled model of an airship at 0, 5, and 10 degrees AoA using CFD [51]. The pressure coefficient is another vital parameter investigated along the axial axis of the aerostat at different AoA and different planes [47, 49, 51, 52, 53]. These pressure coefficients can describe the effect of the control surfaces on the flow separation characteristics of the aerostat well. There are some works that report the investigation of the wall effects of wind tunnels on the experimental results, such as [46, 53]. It is reported that a complete tethered aerostat can be used for CFD simulation analysis, as reported in [54]. In that work, the analysis of the tether force and the balloon displacement for different wind conditions was investigated.

2.3   Mathematical modelling of the tethered aerostat

A mathematical model helps in the quantitative analysis of any dynamic system. Analysing the behaviour of the system under different operating conditions, stability analysis, performance evaluation of the system, and control system development for the system are the major applications of a mathematical model. Tethered aerostats can be found in various applications, such as surveillance, station keeping, imaging, and aerial monitoring, where the mathematical model is used for the design and development of those systems. The applications of the tethered aerostat, and therefore the literature, can be broadly classified as low-altitude applications and high-altitude (stratospheric) applications. There is a predominant difference between the approaches for modelling the tethered aerostat in these two classes. A brief literature survey on the mathematical modelling of the tethered aerostat is presented in the coming sections.

The mathematical model of a tethered aerostat has two parts: the aerostat model and the tether model. A non-linear model of the tether and the TCOM aerostat was presented by Jones, in which a discretised tether model was developed [36]. As the application is at high altitudes, the long tether model includes bending stiffness and damping effects. Jones has presented an aerodynamic model for the high-altitude TCOM aerostat by considering the aerostat as different segments [55]. An extended study presented by Badesha and Jones demonstrated the static and dynamic stability analysis of the aerostat using aerodynamic coefficients [43]. This model involves the experimental data obtained from wind tunnel tests. Kang and Lee presented the necessity of a highly non-linear tether model for a high-altitude application [31]. As the altitude increases, the tether behaviour becomes more dominant in the tethered aerostat response. They have presented a highly non-linear cable model that has a resemblance to the wave equations. There are similar works that have considered a non-linear tether model with elastic properties and stiffness for very high-altitude applications, such as a high-altitude balloon undergoing a thunderstorm [56], a finite element model for a high-altitude tethered balloon [57], and a non-linear tether model with axial stiffness and damping for a high-altitude balloon system [58].

A mathematical model of the low-altitude multi-tethered aerostat for a telescope application was presented by Lambert [37]. A spherical aerostat and a streamline aerostat were used for the analysis and compared for their performance. The aerodynamic model for the aerostat was obtained by the component breakdown method, in which segmentation of the aerostat body was considered. The aerodynamic coefficients were calculated using analytical methods. A detailed study on the position of tether attachment was presented as well. A tether model was presented by Pant et al. for a low-altitude tethered aerostat application [32, 59]. The tether was assumed to have no bending stiffness or elastic properties. They have validated the tether model with experimental test data. Experimental validation of the model proposed by Lambert can be found in [60], in which Pontbriand and Nahon have performed an outdoor experimental test with a tethered spherical aerostat. In all the above-mentioned works, the leashes (harness) connecting the tether and the aerostat were assumed to be a rigid body along with the aerostat. It is justified because most of these works investigated the control problem with tether length control; no effort was reported for controlling the leashes. An exception is a work by Deschesnes and Nahon in which they investigated different approaches to suppress the vibration on the leashes [61]. Even though they investigated some active and passive methods for leash control, a mathematical model was not reported.

Being an aerial vehicle, the aerodynamic part of the model has great importance in the aerostat's dynamics, and by properly developing an aerodynamic model of the aerostat, the design, analysis, and control system for the vehicle can be implemented in an efficient and effective manner. The mathematical modelling of aerostats is similar to that of aircraft in the sense that the dynamics involved in the operation of both vehicles are similar, with some exceptions. The significant difference in the dynamic model of an LTA vehicle from that of other aerial vehicles is the inclusion of the buoyancy effect, the added mass effect, and the tether dynamics. The literature regarding the mathematical modelling of aerostats can be broadly classified based on the method used for obtaining the model parameters as analytical methods, experimental methods, and numerical methods. It can be seen that the aerodynamic modelling of aerostats was presented first in a detailed manner by Munk and then followed by other contributors [62, 63]. The majority of the literature on the aerodynamic model of aerostats is based on the work of Jones and DeLaurier [64]. Some works reported the analytical modelling of the aerodynamics of airships or aerostats, in most of which the model was highly dependent on the experimental data and was not generalisable for a class of vehicles [65], [36], [66], [67].

2.3.1 Stability derivatives

Since the functional form of aerodynamic effects in the form of forces and moments on the aerial vehicle is not easily available for direct measurement, mathematical derivatives are needed to model these forces and moments [68], [69]. Derivatives are a measure of the degree of variation of a function or a parameter with respect to input or another parameter. In the field of dynamics, those derivatives that can represent the sensitivity of the model to some motion variables are termed stability derivatives because these derivatives help simplify the model so that stability analysis and control of the model become easy. Aerodynamic models using stability derivatives are considered an accurate approach to the mathematical modelling of aerial vehicles. Since their introduction by Bryan [70] in 1911, the stability derivatives have been considered the basis for modelling the aerodynamic loads in the equations of motion of the aerodynamic systems. Different applications of stability derivatives include the linearisation of the model for a steady (otherwise with small perturbations) flight to describe the dynamic stability of the vehicle, control system design, trajectory design, aerodynamic shape optimisation, and the design and development of flight simulators [71].

Stability derivatives can be classified as velocity derivatives and acceleration derivatives (added mass terms). The derivatives due to acceleration are usually neglected for the dynamics of heavier-than-air vehicles because of their significantly small values. The importance of the added mass terms in a dynamic system model is based on the relative densities of the body and the fluid [19]. For underwater vehicles and LTA systems (such as aerostats, airships, and hot air balloons), where the added mass effect cannot be neglected, derivatives due to acceleration terms will be included as the added mass and added inertia terms [72]. Added mass, or apparent mass, or virtual mass are the terms that are included in the dynamic model of aerodynamic geometries moving through fluids to represent the forces and torques due to the inertia of the geometry. The second class, the stability derivatives due to the velocity of the motion variables, is usually considered along with the aerodynamic force terms in the dynamic model [73]. In general, these terms are called stability derivatives, and those due to the acceleration terms are called added mass terms.

The estimation or extraction of the stability derivatives was reported mainly using flight tests, wind tunnel tests, and numerical methods. Some of the important works reported in these classes are presented here. The estimation of the aerodynamic derivatives for the LTA vehicles using a semi-empirical procedure was reported in [64]. They have considered the hull-fin interference using an analytical model and the static wind tunnel data, by which the dynamic

17

motion of the vehicle can be well represented. As an extension, they have investigated the longitudinal and lateral stability derivatives of the TCOM 71M aerostat [43]. The static and dynamic stability of the aerostat was also verified by the model obtained using the stability derivatives. A neural network-based prediction of the dynamic derivatives of an aircraft at high AoA was presented by [74]. They provided decent results for pitching moment derivatives compared to the wind tunnel experimental data. A wind tunnel-based stability derivative estimation for a hybrid buoyant aerial vehicle is presented by [75]. They compared the stability of the vehicle with and without wings. Even though the physical realism of the experimental methods, such as flight tests and wind tunnel tests, is highly appreciable, they have some limitations, such as scaling errors, blocking effects, and high costs. A system identification approach for obtaining the stability derivatives of an LTA vehicle undergoing swing oscillation was presented in [76]. Computational methods were also reported for the estimation of stability derivatives of LTA vehicles [77].

Considering the fact that CFD solvers are now capable of handling high-performance tasks due to developments in digital computing, their application in the extraction of stability derivatives is being largely explored by the research community. Most of the limitations of the experimental method, such as the wind tunnel tests, can be overcome by the CFD-based methods. A forced oscillation of the body was used to extract the stability derivatives. The motion variables other than the oscillating parameter are considered to be zero as per the linear theory. A forced periodic oscillation is imposed on the respective degree of freedom (DoF) of the model to calculate the respective dynamic derivative. For example, the moment stability derivative due to the pitch rate can be extracted by imposing a periodic angular oscillation about the lateral axis and centre of the moment [69]. The measured moment response will be used to extract the derivative based on the Fourier series approximation. Such a methodology for the estimation of stability derivatives using CFD for the Zhiyuan airship is presented in [78]. Stability derivatives involved in the pitch and heave motions of the airship were investigated using forced sinusoidal oscillations. They have presented the effect of the tail fins on stability as well. Stability derivatives prediction for a full aircraft configuration was investigated to represent the unsteady aerodynamic load on the flight dynamics for transonic speeds and larger AoA as in [79, 80]. An investigation of the stability of hybrid airships using the analysis of stability

18

derivatives was presented by [81]. The added mass and inertia terms extracted for an underwater vehicle were reported in [82]. They have made a comparison of their results with the results from an experimental procedure and theoretical results. The aerodynamic shape optimisation using stability derivatives addressed by [83] considered the derivatives as constraints in the optimisation problem.

The forced sinusoidal oscillation-based stability derivative extraction is the most popular method due to the well-developed Fourier theory and easy practical realisation [73, 84]. There are also works reported with different methodologies, such as the work reported in [85]. The history effect and the actuation of the body using piece-wise velocity functions were the basis of their methodology. Even though the rotational dynamics were not able to be captured, the methodology was validated with existing data. A detailed literature review of the existing methodologies for stability derivative extraction using CFD can be found in [86].

2.4   Research gaps identified

A detailed literature survey on the aerostat-based systems, their operation and applications, aerodynamic analysis of aerostats under the influence of wind, and the mathematical modelling of the tethered aerostats was carried out. In this section, the significant research gaps and the open challenges identified in the scope of the current research are presented.

- **Aerodynamic analysis under unsteady conditions.** Even though there is a sufficient volume of work reported on the aerodynamic analysis of the aerostat under steady wind conditions, a proper computational analysis under unsteady wind conditions was not reported. As the analysis involves responses due to changes in a single parameter, it would be easy to have a generalised form for the individual family of responses. It will help in the analysis as well as the prediction of similar responses. A generalisation method or a scaling method for similar aerodynamic responses has not been reported in the open literature.

- **Aerostat stability derivatives for steady conditions.** Even though the investigation of stability derivatives of an aerostat under steady flow conditions using CFD is reported, it is not done for the complete set of derivatives required for a mathematical model of the aerostat.

- **Aerostat stability derivatives for unsteady conditions.** A study on the stability derivatives of an aerostat under unsteady flow conditions or unsteady motion conditions is not reported in the open literature.

- **Effect of shape of the aerostat on the stability derivatives.** Since the aerodynamic shape of the aerostat hull plays an essential role in the stability dynamics, its effect will definitely reflect on the stability derivatives as well. There is no work that reports the analysis of the impact of aerostat shape on the stability derivatives.

2.5 Formulation of objectives for the current research

The research problem can be stated as '**To analyse the aerodynamic performance of the aerostat under unsteady wind conditions and investigate the mathematical model of the tethered aerostat.**'

Based on the literature survey conducted, the following research objectives are formulated for the current research:

1. **To analyse the aerodynamic parameters of an aerostat under unsteady wind conditions.**

2. **To investigate the stability derivatives of an aerostat under steady wind conditions.**

3. **To investigate the stability derivatives of an aerostat under unsteady wind and motion conditions.**

4. **To investigate the decoupled longitudinal model for the single-tethered aerostat using stability derivatives.**

2.6 Summary

This chapter presented a detailed literature review on aerostats, aerodynamic analysis, mathematical modelling, and stability derivatives of the aerostat. A set of research gaps were identified and listed, which guided the author to formulate the research objectives of the current research. The research gaps include the unsteady analysis of the aerostat, scaling of the aerostat responses, stability derivative extraction under steady and unsteady conditions, the effect of the aerostat shape on the stability derivatives, and the mathematical modelling of the aerostat using stability derivatives. Four research objectives were formulated based on these research gaps, as given in Section 2.5.

Chapter 3

Research Methodology

3.1 Introduction

This section presents the research methodology adopted for the current research. Various models considered part of the methodology, such as the aerostat geometry and wind gust models, are presented. The simulation setup used for the current research is briefly explained. As the stability derivative extraction of the aerostat is an integral part of the research, a detailed discussion of the adopted methodology is presented. A set of results supporting the validation of the simulation setup used for the current research is also presented at the end.

3.2 Aerostat geometry

The geometry used for the aerostat 3D model is presented in this section. The current research consists of an aerodynamic analysis part and a stability derivative extraction part using CFD analysis. For the aerodynamic analysis, the Zhiyuan aerostat is used, and for the stability derivative extraction part, four aerostats are used, namely, Zhiyuan, GNVR, NPL, and HAA.

3.2.1 For the aerodynamic analysis

The numerical analysis of the aerostat hull with four tail fins is considered in this work. The standard optimised aerostat hull shapes available in the literature include NPL [87], Wang [88], GNVR [89], and Zhiyuan [33]. By considering the low drag profile and the availability of extensive experimental data, the Zhiyuan profile is selected as the base case for the current research. In this research, a four-winged aerostat with a '+' orientation of the wings is considered. The length of the aerostat is fixed at 13.5 m by considering the required volume for the desired payload weight. The scaled values of the geometric parameters of the aerostat used in the current research are given in Table 3.1, along with the actual aerostat and the scaled test aerostat model.

Table 3.1: Zhiyuan aerostat geometric parameters.

Parameter	Actual Zhiyuan [33]	Test model [33]	Current model
Volume (m^3)	750	0.29	118
Diameter (m)	7.58	0.56	4.09
Length (m)	25	1.8	13.5
Surface area (m^2)	480.4	2.57	140
Reference area (m^2)	82.5	0.44	24
Reference length (m)	9.1	0.66	4.9
Reynolds number	$(1.8 - 9.3)e^6$	$3.2e^6$	$4.6e^6$

The 2D cross-section of the Zhiyuan aerostat hull geometry, shown in Figure 3.1 (a), was extracted from the 2D profile reported in [33]. It is the scaled version of the 'Actual Zhiyuan' and was used as a 'Test model' for the experimental data collection reported in [33]. The length scaling factor of the aerostat model used for the current research ('Current model') is 1.85 with respect to the actual aerostat. This is selected according to the payload carrying capacity of 5 kg needed for the proposed application. The model has a length of 13.5 m, a maximum diameter of 4.09 m and a volume of 118 m^3. There are some modifications done to the Zhiyuan geometry to develop the 3D model. First, the gondola is removed from the envelope because it is not required for the current application. Second, the wings of the Zhiyuan model are slightly modified to simplify the aerostat structure. The wing base section is modified to be completely attached to the aerostat hull. The wing geometry is simplified without losing the dynamic characteristics of the aerostat. The 3D model of the aerostat is shown in Figures 3.1 (b) and (c). The experimental results obtained for the 'Test model' are used for the validation studies in the current research.

In this research, the characteristic length is used to define the Reynolds number. The Reynolds number is defined as,

$$Re = \frac{\rho V_0 L_A}{\mu}$$

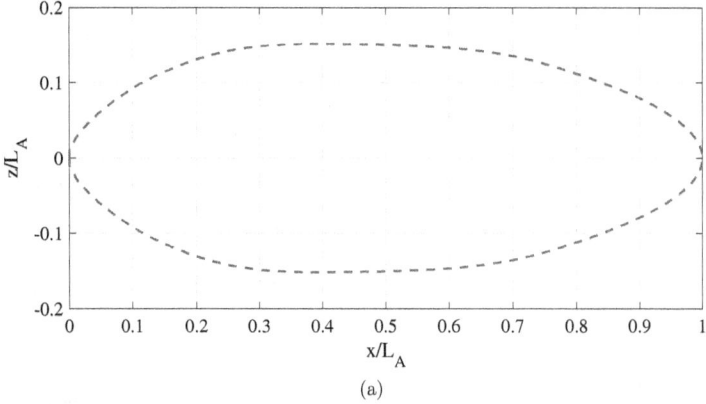

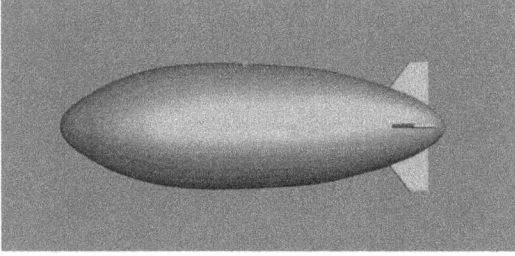

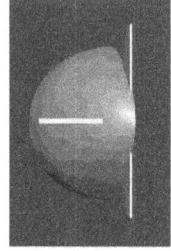

Figure 3.1: (a) Zhiyuan aerostat 2D profile, (b) 3D profile - side view, and (c) rear view.

where ρ is the fluid density, V_0 is the mean flow velocity, L_A is the characteristic length, and μ is the dynamic viscosity. $V^{1/3}$ is used as the characteristic length in this research, where V is the aerostat envelope volume. Thus, the expression for Reynolds number becomes,

$$Re = \frac{\rho V_0 V^{1/3}}{\mu}$$

This characteristic length is used as the reference value for length in the simulation, and $V^{2/3}$ as the reference value for the area.

3.2.2 For the stability derivative analysis

There are several aerostat/airship geometries available in the open literature. The NPL shape has been designed and developed by the National Physical Laboratory to reduce the chances of flow separation [90]. It was achieved by providing a continuous variation in the

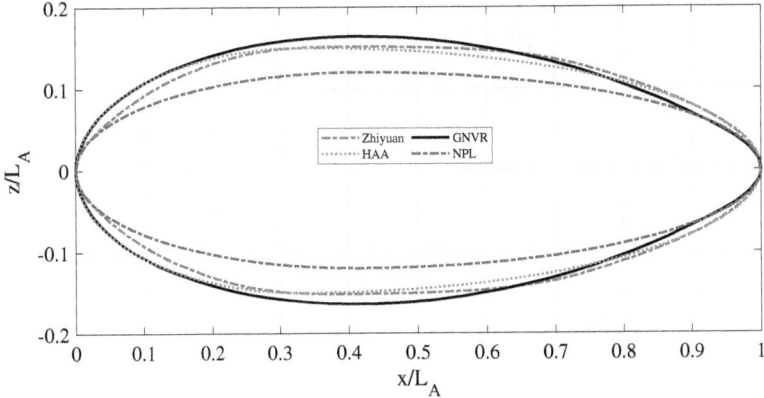

Figure 3.2: Aerostat profiles considered for the stability derivative analysis.

radius of curvature of the profile. The GNVR shape has been reported to have a low drag profile [90]. Both of these shapes were able to represent a combination of different simple geometries, such as ellipses and parabolas. It made them feasible for geometry optimisation. A high-altitude application was reported, which was claimed to have a minimum weight to increase the payload and stability (HAA shape) [91]. All the above-mentioned shapes were proposed for stratospheric operations. A bionic design for the airship has been reported for the minimisation of drag (bionic shape) [89]. It was achieved by imitating the morphological features of an aquatic animal (*Physalia physalis*). Other low drag profiles were also reported for certain applications, such as the Zhiyuan shape [78] and Wang shape [92].

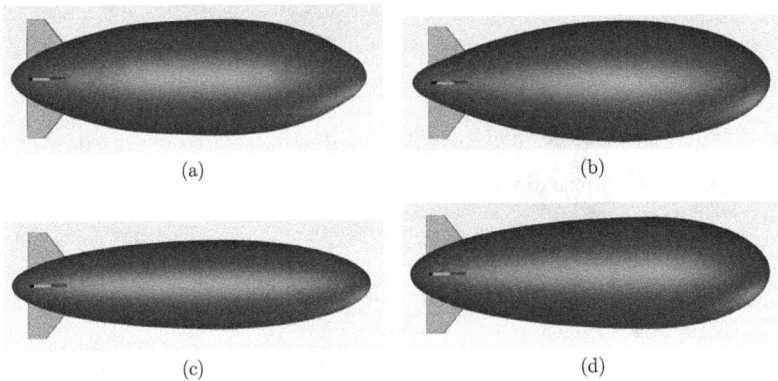

Figure 3.3: Aerostat 3D models; (a) Zhiyuan, (b) GNVR, (c) NPL, and (d) HAA shape.

Table 3.2: Geometric parameters of the four aerostats (scaled for the current research).

Parameter	Zhiyuan	GNVR	NPL	HAA
Volume (m^3)	118	130	110	115
Diameter (m)	4.09	4.55	3.33	4.09
Length (m)	13.5	13.5	13.5	13.5
Reference area (m^2)	24	25.6	22.9	23.6
Reference length (m)	4.9	5.06	4.79	4.86
Reynolds number	$(1.8-9.3)e^6$	$(1.8-9.3)e^6$	$(1.8-9.3)e^6$	$(1.8-9.3)e^6$

For the stability derivative extraction analysis, four aerostat envelope shapes are considered for the comparison of stability derivatives. They are Zhiyuan shape [78], GNVR shape [90], NPL shape [90] and HAA shape [91]. The 2D coordinates of the four aerostat shapes are extracted from the geometric profile data reported in the above-mentioned literature. The 2D profiles for the four aerostats are shown in Figure 3.2. These shapes were selected by considering the variation of their leading-edge geometry, diameter, and the lift-drag characteristics. The 3D model of the four aerostats used for the simulation analysis is shown in Figure 3.3. The maximum diameter is 4.09 m for the Zhiyuan shape, 4.55 m for the GNVR shape, 3.33 m for the NPL shape and 4.09 m for the HAA shape. Among the four aerostat shapes considered, GNVR has the maximum volume, and NPL has the least volume. Table 3.2 consolidates the geometric parameters of the four aerostats. The scaling criterion used for the Zhiyuan shape is used for the other three aerostats as well.

3.3 Wind gust models

The unsteady flow conditions are induced by including wind gusts in the free stream velocity. This section presents the details of various wind gust models considered for the current research. In this research, the wind along the axial axis alone is considered. Therefore, a head-on gust along the X-axis of the aerostat is considered. For the aerodynamic analysis of the aerostat, five different discrete gust models are considered. They are the International Electrotechnical Commission (IEC) gust, sinusoidal gust, cosine gust, and two modified gust models to incorporate asymmetry in their profiles. A set of continuous sinusoidal wind gust models is used for the stability derivative extraction. The occurrence of continuous wind gusts is frequently compared to discrete gusts, and that is the reason for selecting the continuous wind gust for the stability derivative analysis [93, 94].

3.3.1 IEC gust (Gust 1)

The Extreme Operating Gust (EOG), described in the IEC standard [95], is used in this work for the aerodynamic analysis of the aerostat. The aforementioned standard is specified for small wind turbines that are operating at similar operating conditions as aerostats, such as altitude and wind conditions. The IEC gust velocity profile can be expressed as:

$$V(t) = \begin{cases} V_m - 0.37V_g sin(3\pi t/T_0)(1 - cos(2\pi t/T_0)) & :0 \le t \le T_0 \\ V_m & :\text{otherwise} \end{cases} \tag{3.1}$$

where V_m is the mean free stream velocity, V_g is the gust amplitude, and T_0 is the gust time period. The velocity profile based on Equation (3.1) is shown in Figure 3.4 (a) for a mean wind velocity of 7.5 m/s, gust amplitude of 12.5 m/s, and a time period of 10.5 seconds. The gust amplitude is selected as 70% of the mean wind velocity. The large time period of the gust is selected to adequately capture the variations in the aerodynamic parameters of the aerostat. The X-axis (time) is normalised by dividing the time by the time period of the gust, as shown below.

$$\tau = \frac{t}{T} \tag{3.2}$$

The mean wind speed is taken from the NIWE wind data from the Gujarat coastal area at an altitude of 100 m above sea level [96]. The amplitude of the gust is extracted from the same wind data based on the following formulations [97]:

$$V_g = max\{(V_{max} - V_{min})T_g\} \tag{3.3}$$

where T_g is the constant number of wind data points (a data window) that will be moved forward from the first data point to the last. The difference between the maximum and minimum values of the velocity will be calculated as the window progresses through the entire data. The maximum among these differences is selected as the amplitude of the gust.

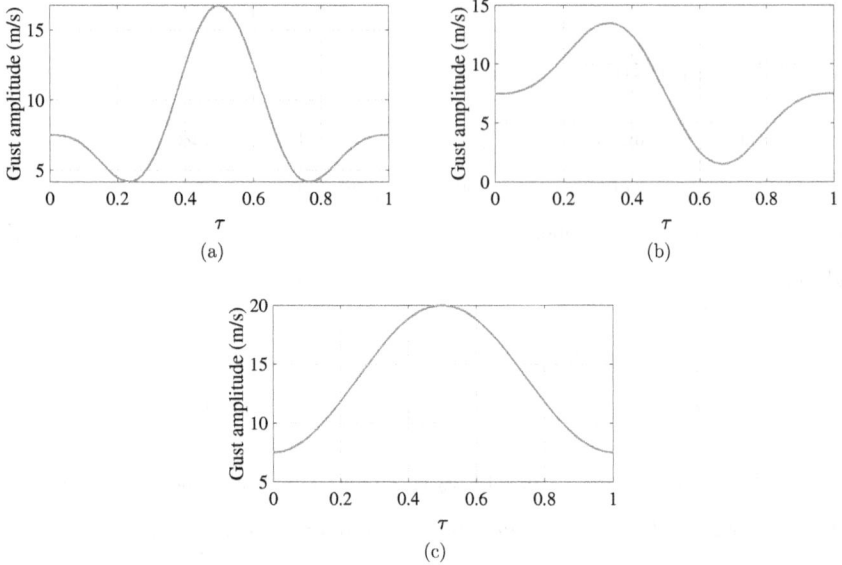

Figure 3.4: (a) Extreme Operating Gust (EOG), (b) sinusoidal gust, and (c) 1-cosine gust.

3.3.2 Sinusoidal gust (Gust 2)

The second wind gust model considered for the aerodynamic analysis of the aerostat is the sinusoidal gust. The gust model is modified slightly to remove the discontinuity at the beginning and end. A more realistic and smooth sinusoidal nature for the gust is obtained by modifying the IEC gust formulation.

The mathematical expression for the presented sinusoidal wind gust can be given as,

$$V(t) = V_m + 0.37V_g \sin(\frac{2\pi t}{T_0})(1 - \cos(\frac{2\pi t}{T_0})) \tag{3.4}$$

where the parameters V_m, V_g, and T_0 are the mean wind velocity 7.5 m/s, gust amplitude 12.5 m/s, and gust duration 10.5 seconds, respectively.

The velocity profile of the sinusoidal gust is shown in Figure 3.4 (b).

3.3.3 Cosine gust (Gust 3)

A standard 1-cosine gust model is the third wind gust model used for the analysis. By applying the magnitude of 12.5 m/s, this gust model has the maximum amplitude of the five

28

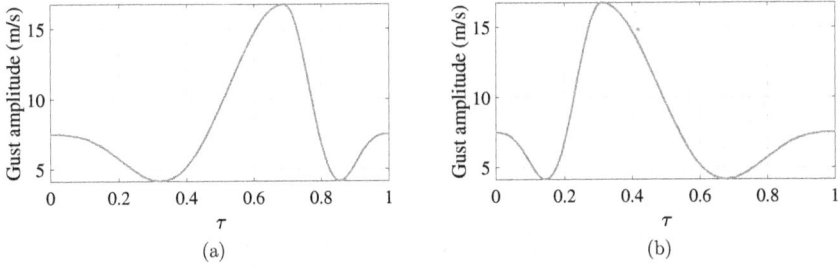

Figure 3.5: Asymmetric gust models: (a) gust 4 and (b) gust 5.

wind gust models.

The mathematical expression for the presented 1-cosine wind gust can be given as,

$$V(t) = V_m + (V_g/2)(1 - \cos(\frac{2\pi t}{T_0})) \tag{3.5}$$

where the parameters V_m, V_g, and T_0 are the mean wind velocity, 7.5 m/s, gust amplitude, 12.5 m/s, and gust duration, 10.5 seconds, respectively.

The velocity profile of the 1-cosine gust is shown in Figure 3.4 (c).

3.3.4   Modified IEC gust models (Gust 4 and Gust 5)

The three gust models considered so far are symmetrical in nature. To analyse the effect of asymmetry in the wind gust model, two asymmetric wind gusts are considered. They are obtained by modifying the IEC wind gust model. The parameters (such as the time periods) used for getting the velocity profiles for gusts 4 and 5 are selected in such a way that the effect of gust amplitude growth is included.

The mathematical expression for the presented asymmetric wind gust model with a large rising time and a short falling time can be given as,

$$V(t) = \begin{cases} V_m - 0.37V_g \sin(\frac{3\pi t}{14.5})(1 - \cos(\frac{2\pi t}{14.5})) & : \ 0 < t < (14.5/2) \\ V_m - 0.37V_g \sin(\frac{3\pi(t-4)}{6.5})(1 - \cos(\frac{2\pi(t-4)}{6.5})) & : \ (14.5/2) < x < T_0 \end{cases} \tag{3.6}$$

The mathematical expression for the presented asymmetric wind gust model with a small rising

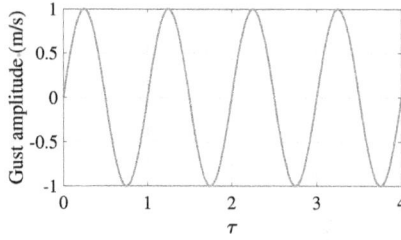

Figure 3.6: Continuous sinusoidal wind gust.

time and large falling time can be given as,

$$
V(t) = \begin{cases} V_m - 0.37 V_g \sin(\frac{3\pi t}{6.5})(1 - \cos(\frac{2\pi t}{6.5})) & : \; 0 < t < (6.5/2) \\ V_m - 0.37 V_g \sin(\frac{3\pi(t+4)}{14.5})(1 - \cos(\frac{2\pi(t+4)}{14.5})) & : \; (6.5/2) < x < T_0 \end{cases} \tag{3.7}
$$

where the parameters V_m, V_g, and T_0 are the mean wind velocity, 7.5 m/s, gust amplitude, 12.5 m/s, and gust duration, 10.5 seconds, respectively.

The velocity profile of the two asymmetric wind gust models is shown in Figure 3.5.

3.3.5 Continuous sinusoidal wind gust model

For the stability derivative extraction of the aerostat, a set of continuous wind gust models is considered. A sinusoidal gust is considered, which is different from the one presented in Section 3.3.2. Since the gust considered is of a continuous nature, the problem of discontinuity at the beginning and end of the gust does not occur. The amplitude of the mean wind is taken as 7.5 m/s here as well. Instead of considering a large gust amplitude, a smaller amplitude is considered because the stability derivative extraction methodology demands a small amplitude oscillation.

The mathematical expression for the continuous sinusoidal gust model is given as follows,

$$
V(t) = V_g \sin(\omega t) \tag{3.8}
$$

The velocity profile of the continuous sinusoidal wind gust model is shown in Figure 3.6. The gust amplitude is taken as 1 m/s for the surge and heave motions.

3.4 Numerical simulation setup

3.4.1 Governing equations

All the simulations reported in this research were performed using the commercial CFD code Fluent 19.0 [98]. The current research involves a 3D, incompressible, and unsteady flow, which will be solved using the Reynolds Averaged Navier Stokes (RANS) equations [98]. Even though the RANS solver is not preferable for higher angles of attack conditions, it is the most widely used solver in the industry. The governing equations are the mass conservation equation and the momentum conservation equation, as given in Equations (3.9) and (3.10), respectively.

$$\nabla \cdot \vec{\mathbf{V}} = 0 \tag{3.9}$$

$$\rho \frac{\partial \vec{V}}{\partial t} + \rho \nabla \cdot \left(\vec{\mathbf{V}}\vec{\mathbf{V}} \right) = -\nabla p + \nabla \cdot \left(\bar{\bar{\tau}} \right) + \rho \vec{g} \tag{3.10}$$

where ρ is the density, $\vec{\mathbf{V}}$ is the velocity, p is the pressure, and $\vec{g}$ is the acceleration due to gravity. The stress tensor $\bar{\bar{\tau}}$ can be expressed as,

$$\bar{\bar{\tau}} = (\boldsymbol{\mu} + \boldsymbol{\mu}_t) \left[\nabla \vec{\mathbf{V}} + \left(\vec{\mathbf{V}} \right)^T - \frac{2}{3} \nabla \cdot \vec{\mathbf{V}} \mathbf{I} \right]$$

where $\boldsymbol{\mu}$ is the dynamic viscosity, $\boldsymbol{\mu}_t$ is the turbulent viscosity, and $\mathbf{I}$ is an identity matrix.

The pressure and velocity equations are coupled using the SIMPLE algorithm [98]. The spatial discretisation of the convective terms is taken care of by the second-order upwind scheme. The values of residuals for continuity, velocity, and turbulence quantities were set to the order of 10^{-5}.

The realizable $k - \epsilon$ model, which is a member of the two-equation model of the RANS turbulence model, is used for the simulation. The transport equations for the realizable $k - \epsilon$ model are given in Equations (3.11) and (3.12).

$$\frac{\partial}{\partial t}(\rho k) + \frac{\partial}{\partial x_j}(\rho k u_j) = \frac{\partial}{\partial x_j}\left[\left(\mu + \frac{\mu_t}{\sigma_k} \right) \frac{\partial k}{\partial x_j} \right] + G_K + G_b - \rho \epsilon - Y_M + S_K \tag{3.11}$$

31

$$\frac{\partial}{\partial t}(\rho\epsilon) + \frac{\partial}{\partial x_j}(\rho\epsilon u_j) = \frac{\partial}{\partial x_j}\left[\left(\mu + \frac{\mu_t}{\sigma_\epsilon}\right)\frac{\partial\epsilon}{\partial x_j}\right] + \rho C_1 S\epsilon - \rho C_2 \frac{\epsilon^2}{k + \sqrt{v\epsilon}} +$$
$$C_{1\epsilon}\frac{\epsilon}{k}C_{3\epsilon}G_b + S_\epsilon$$

(3.12)

where G_k and G_b are the turbulent kinetic energy due to mean velocity gradient and buoyancy, respectively; Y_M is the dissipation due to fluctuating dilation in compressible turbulence; C_2 and $C_{1\epsilon}$ are constants; σ_k and σ_ϵ are the turbulent Prandtl numbers for k and ϵ respectively; and S_k and S_ϵ are user-defined source terms.

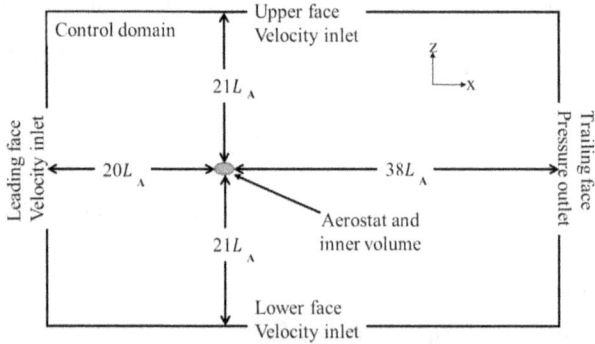

(a)

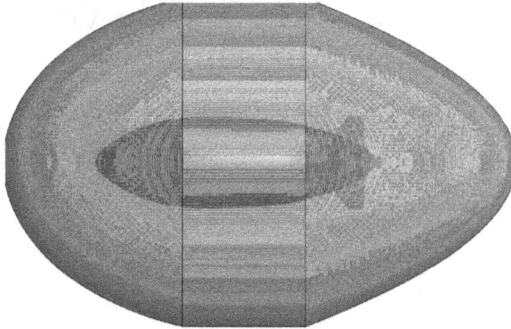

(b)

Figure 3.7: (a) Computational domain used for the aerodynamic analysis and (b) computational grid close to the volume.

3.4.2 Computational grid and domain

The simulations performed for the current research fall under two different sets: one used for the aerodynamic analysis of the aerostat under wind gusts, and the second one used for the stability derivative extraction. The computational grid and domain used for these two sets of analyses are slightly different.

For the aerodynamic analysis

The computational domain around the aerostat used for the aerodynamic characteristic analysis is considered a rectangular volume, as shown in Figure 3.7 (a). The dimension of the rectangular volume is $60L_A \times 43L_A \times 21L_A$, where L_A is the length of the aerostat model. The leading face of the rectangular volume is kept at a distance of $20L_A$ ahead of the leading edge of the aerostat, and the upper and lower faces are at a distance of about $21L_A$ from the aerostat. A volume closer to the aerostat is considered, as shown in Figure 3.7 (b), to capture the flow dynamics in the close vicinity of the model precisely. This inner volume has a length of $2L_A$ and a maximum diameter of $1.2L_A$.

The aerostat surface is kept as a stationary wall with no-slip boundary conditions. The leading face, top face, bottom face, and front face of the rectangular volume are kept as velocity inlet, the trailing face as pressure outlet and the rear face is kept as a symmetry plane. The boundary conditions are shown in Figure 3.7 (a).

The aerostat simulations were done using the cut-cell grid generated using ANSYS Fluent Meshing. The advantages of using such a meshing topology are easy implementation, adaptive meshing capability, and faster convergence. The outer and inner volumes around the aerostat are meshed using 7 and 2 refinements, respectively. The initial grid specified for the outer volume consists of $160 \times 100 \times 50$ cells, and the inner volume consists of $38 \times 28 \times 14$ cells along the x, y, and z axes. The size of the cells on the outermost surface of the outer volume is $19.74\ m \times 19.28\ m \times 19.28\ m$. The cell size is halved for each of the refinements until the outer surface of the inner volume is reached. The cell size on the outer surface of the inner volume is $0.154\ m \times 0.15\ m \times 0.15\ m$. The number of diffusion layers specified for these refinements is 8. This initial grid is further modified automatically to adapt to the geometry. The boundary layer is specified as 22 layers with a growth rate of 1.35 and an initial cell size of $5e^{-05}$. With

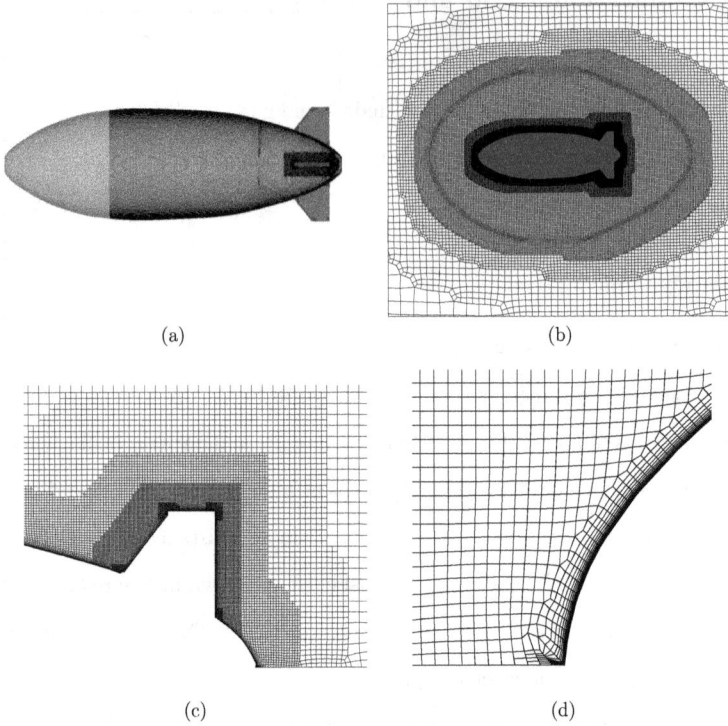

(a)   (b)

(c)   (d)

Figure 3.8: Computational grid used for the aerodynamic analysis: (a) surface grid, (b) close volume, (c) wing section, and (d) leading edge.

the above specified grid configuration, the average value of the wall 'Y+' is found to be 0.435 with a minimum of 0.0349 and a maximum of 1.

Figure 3.8 shows the final grid configuration done for the aerostat. Figure 3.8 (a) shows the grid on the aerostat surface. It is clear from the figure that the adaptation to the geometry has well resolved the curvature of the aerostat. Figure 3.8 (b) shows a close view of the aerostat and the control domain close to the aerostat surface. Figure 3.8 (c) shows a close view of the grid at the aerostat wing section. It clearly shows that the wings and curvature are very well resolved with fine meshing. Figure 3.8 (d) shows a close view of the aerostat's leading edge to show the boundary layer.

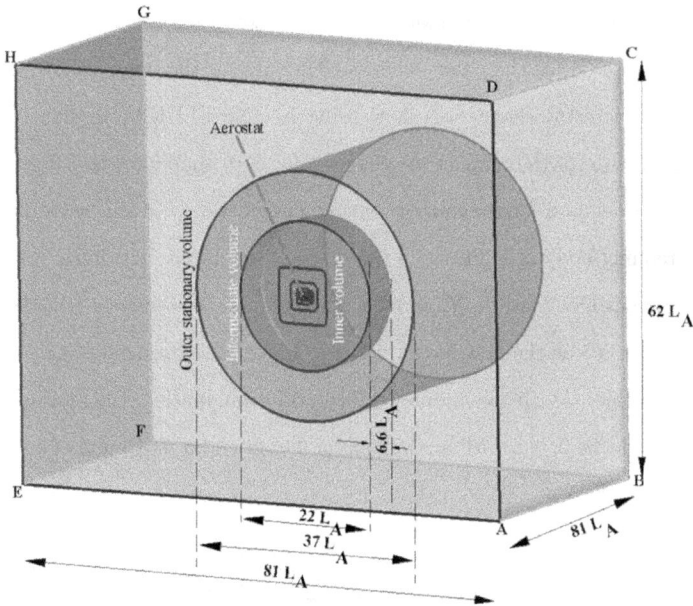

Figure 3.9: Computational domain used for the stability derivative analysis.

For the stability derivative analysis

The computational domain used for the simulation is shown in Figure 3.9. The dimensions of the domain are selected after an initial domain independence study. On each of the four faces of the outermost box volume depicted in Figure 3.9, there are four velocity inlets (faces ABCD, BCFG, CDHG, and ABEF), symmetry on one face (face ADHE), and a pressure outlet on the other (face EFGH). The outer and intermediate volumes were common for all the aerostat simulations presented in this study. The dimensions of the domain are specified as a function of the aerostat model length. The dynamic mesh feature of ANSYS Fluent [98] realised the heave and surge motion simulations, and the pitching motion was realised by the grid motion feature. The inner and intermediate volumes were specified as rigid bodies, and the outer volume was kept stationary for the surge and heave motions. The inner volume was forced to move by the specified user-defined function, and the outer and intermediate volumes were kept stationary for the pitching motion simulation.

The computational grid used for the current research was made using GAMBIT and FLU-ENT meshing. The grid for outer and intermediate volumes (refer to Figure 3.9) was done using GAMBIT with a total of 0.1 million tetrahedral-hexahedral hybrid cells. Comparatively, the coarse grid was used for the outer volumes. The grid for the inner volume with aerostat (different for each aerostat shape) was done using ANSYS FLUENT meshing, as shown in Figure 3.10. The range of the number of cells used for each shape is 3 to 3.5 million. A cut-cell grid topology was used for the inner volumes. The proximity of the aerostat was meshed finely using a refining body of influence in the shape of an ellipse. The inner volume grid has $50 \times 50 \times 15$ cells along the X, Y, and Z cartesian axes. The viscous boundary layer was specified with $5E - 005$ as the first layer thickness, 1.35 as the stretching ratio, and 18 layers for the aerostat surface. With the above-specified grid configuration, the average value of the wall 'Y+' is found to be 0.435, with a minimum of 0.0349 and a maximum of 1.

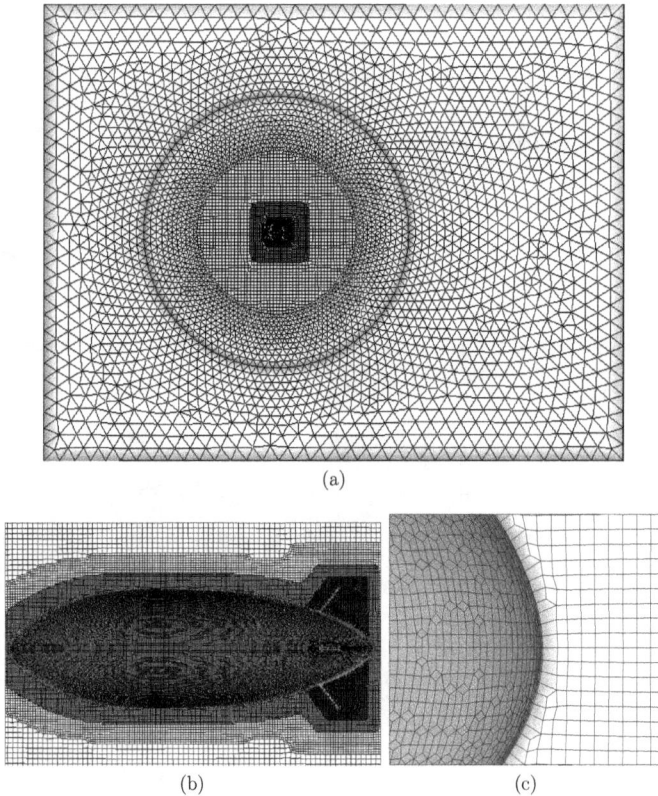

(a)

(b)

(c)

Figure 3.10: Computational grid used for the stability derivative analysis: (a) control domain, (b) surface grid, and (c) leading edge.

3.5 Stability derivative extraction

In this research, the longitudinal model of the aerostat is considered, and the motion of the aerostat in its longitudinal axes is considered. The lateral motion of the aerostat is assumed to be negated by the control surfaces. The motion variables involved in the decoupled linear longitudinal dynamic model are u, w, and q.

The longitudinal model of the aerostat can be expressed in matrix form as given in Equation (3.13). A detailed derivation of this model is given in Chapter 6. The terms with dots are the stability derivatives involved in the longitudinal dynamics of the aerostat. Table 3.3 shows the list of the stability derivatives involved in Equation (3.13), along with their interpretation.

$$
\begin{bmatrix} m - \dot{D}_{\dot{u}} & 0 & ma_z - \dot{D}_{\dot{q}} \\ 0 & m - \dot{L}_{\dot{w}} & ma_x + \dot{L}_{\dot{q}} \\ ma_z - \dot{M}_{\dot{u}} & ma_x + \dot{M}_{\dot{w}} & I_y - \dot{M}_{\dot{q}} \end{bmatrix} \begin{bmatrix} \dot{u} \\ \dot{w} \\ \dot{q} \end{bmatrix} = \begin{bmatrix} D_e \\ L_e \\ M_e \end{bmatrix} +
$$
$$
\begin{bmatrix} \dot{D}_u & \dot{D}_w & \dot{D}_q \\ \dot{L}_u & \dot{L}_w & \dot{L}_q \\ \dot{M}_u & \dot{M}_w & \dot{M}_q - ma_x U_e - ma_z W_e \end{bmatrix} \begin{bmatrix} u \\ w \\ q \end{bmatrix} + F_{ext}
$$
(3.13)

The model variables included in the longitudinal model of the aerostat are u, w, q, D, L, and M. These longitudinal dynamic variables are predominant in the heave, surge, and pitch motions of the aerostat. Figure 3.11 (a) shows the orientation of the aerostat axis, wind direction, and the motions considered in this research.

The methodology used for the extraction of the stability derivatives includes the simulation of the model for steady-state, then transient, and finally the oscillations in the respective directions. The forces and moment (D, L and M) are measured for the oscillated motions. A stable cycle of data will be used from these measured responses to extract the stability derivatives using the Fourier series based method discussed in the following sections.

Table 3.3: Stability derivatives for the longitudinal dynamic model of the aerostat.

Symbol	Description
D_e	Static drag derivative
L_e	Static lift derivative
M_e	Static moment derivative
Stability derivatives due to velocity	
$\dot{D}_u$	Drag force derivative due to axial velocity
$\dot{D}_w$	Drag force derivative due to vertical velocity
$\dot{D}_q$	Drag force derivative due to pitch angular velocity
$\dot{L}_u$	Lift force derivative due to axial velocity
$\dot{L}_w$	Lift force derivative due to vertical velocity
$\dot{L}_q$	Lift force derivative due to pitch angular velocity
$\dot{M}_u$	Pitching moment derivative due to axial velocity
$\dot{M}_w$	Pitching moment derivative due to vertical velocity
$\dot{M}_q$	Pitching moment derivative due to pitch angular velocity
Stability derivatives due to acceleration	
$\dot{D}_{\dot{u}}$	Drag force derivative due to axial acceleration
$\dot{D}_{\dot{q}}$	Drag force derivative due to pitch angular
$\dot{L}_{\dot{w}}$	Lift force derivative due to vertical acceleration
$\dot{L}_{\dot{q}}$	Lift force derivative due to pitch angular acceleration
$\dot{M}_{\dot{u}}$	Pitching moment derivative due to axial acceleration
$\dot{M}_{\dot{w}}$	Pitching moment derivative due to vertical acceleration
$\dot{M}_{\dot{q}}$	Pitching moment derivative due to pitch angular acceleration

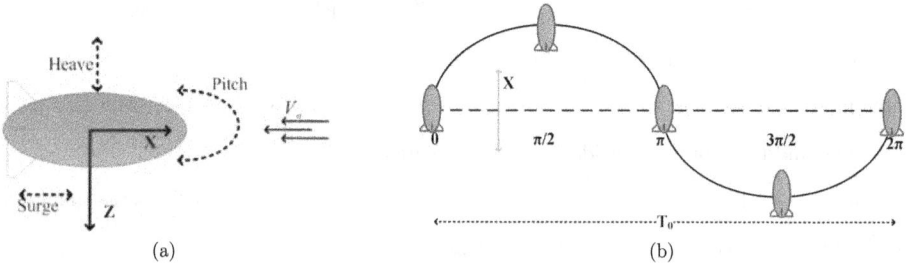

Figure 3.11: (a) Aerostat reference axes and (b) surge motion.

3.5.1 Pure surge oscillations

The stability derivatives associated with the translational forward motion (along the X-axis) can be expressed as the linearly decomposed forces and moment as follows,

$$D_a^s = D_e + \dot{D}_u u + \dot{D}_{\dot{u}} \dot{u}$$
$$L_a^s = L_e + \dot{L}_u u + \dot{L}_{\dot{u}} \dot{u} \tag{3.14}$$
$$M_a^s = M_e + \dot{M}_u u + \dot{M}_{\dot{u}} \dot{u}$$

A forced sinusoidal oscillatory motion will be induced in the aerostat axial direction, as shown in Figure 3.11 (b). The motion can be defined by the displacement $x = u_0 \sin \omega_s t$, velocity $u = \omega_s u_0 \cos \omega_s t$, and acceleration $\dot{u} = -\omega_s^2 u_0 \sin \omega_s t$. Substituting the values of u and $\dot{u}$ into Equation (3.14) will result in,

$$D_a^s = D_e + \dot{D}_u \omega_s u_0 \cos \omega_s t - \dot{D}_{\dot{u}} \omega_s^2 u_0 \sin \omega_s t$$
$$L_a^s = L_e + \dot{L}_u \omega_s u_0 \cos \omega_s t - \dot{L}_{\dot{u}} \omega_s^2 u_0 \sin \omega_s t \tag{3.15}$$
$$M_a^s = M_e + \dot{M}_u \omega_s u_0 \cos \omega_s t - \dot{M}_{\dot{u}} \omega_s^2 u_0 \sin \omega_s t$$

As Equation (3.15) comprises the sine and cosine components, the coefficients can be compared with the corresponding Fourier series representations of the forces and moment with higher-order terms neglected as given in Equation (3.16).

$$D_a = a_0/2 + a_1 \cos \omega_s t + a_2 \sin \omega_s t$$
$$L_a = b_0/2 + b_1 \cos \omega_s t + b_2 \sin \omega_s t \tag{3.16}$$
$$M_a = c_0/2 + c_1 \cos \omega_s t + c_2 \sin \omega_s t$$

By comparing the coefficients of in-phase and out-of-phase components of Equation (3.16) and Equation (3.15),

$$\dot{D}_u = a_1/(\omega_s u_0); \qquad \dot{L}_u = b_1/(\omega_s u_0)$$
$$\dot{M}_u = c_1/(\omega_s u_0); \qquad \dot{D}_{\dot{u}} = -a_2/(\omega_s^2 u_0) \tag{3.17}$$
$$\dot{L}_{\dot{u}} = -b_2/(\omega_s^2 u_0); \qquad \dot{M}_{\dot{u}} = -c_2/(\omega_s^2 u_0)$$

where,

$$a_1 = \frac{2}{T_0} \int_{-T_0/2}^{T_0/2} D_a(t) \cos(\omega_s t) dt \qquad a_2 = \frac{2}{T_0} \int_{-T_0/2}^{T_0/2} D_a(t) \sin(\omega_s t) dt$$

$$b_1 = \frac{2}{T_0} \int_{-T_0/2}^{T_0/2} L_a(t) \cos(\omega_s t) dt \qquad b_2 = \frac{2}{T_0} \int_{-T_0/2}^{T_0/2} L_a(t) \sin(\omega_s t) dt$$

$$c_1 = \frac{2}{T_0} \int_{-T_0/2}^{T_0/2} M_a(t) \cos(\omega_s t) dt \qquad c_2 = \frac{2}{T_0} \int_{-T_0/2}^{T_0/2} M_a(t) \sin(\omega_s t) dt$$

According to Equation (3.17), a full-cycle response of drag, lift, and moment for the forced oscillation along the axial direction will give the six stability derivatives due to axial velocity and acceleration.

3.5.2 Pure heave oscillations

The stability derivatives associated with the translational vertical motion (along the Z-axis) can be expressed as the linearly decomposed forces and moment as follows,

$$D_a^h = D_e + \dot{D}_w w + \dot{D}_{\dot{w}} \dot{w}$$
$$L_a^h = L_e + \dot{L}_w w + \dot{L}_{\dot{w}} \dot{w} \qquad (3.18)$$
$$M_a^h = M_e + \dot{M}_w w + \dot{M}_{\dot{w}} \dot{w}$$

A forced sinusoidal oscillatory motion will be induced in the aerostat vertical direction, as shown in Figure 3.12 (a). The motion can be defined by the displacement $y = w_0 \sin \omega_h t$, velocity $w = \omega_h w_0 \cos \omega_h t$, and acceleration $\dot{w} = -\omega_h^2 w_0 \sin \omega_h t$. Substituting the values of w and $\dot{w}$ into Equation (3.18) will result in,

$$D_a^h = D_e + \dot{D}_w \omega_h w_0 \cos \omega_h t - \dot{D}_{\dot{w}} \omega_h^2 w_0 \sin \omega_h t$$
$$L_a^h = L_e + \dot{L}_w \omega_h w_0 \cos \omega_h t - \dot{L}_{\dot{w}} \omega_h^2 w_0 \sin \omega_h t \qquad (3.19)$$
$$M_a^h = M_e + \dot{M}_w \omega_h w_0 \cos \omega_h t - \dot{M}_{\dot{w}} \omega_h^2 w_0 \sin \omega_h t$$

By comparing the coefficients of in-phase and out-of-phase components of Equation (3.16) and Equation (3.19),

$$\dot{D}_w = a_1/(\omega_h w_0); \qquad \dot{L}_w = b_1/(\omega_h w_0)$$
$$\dot{M}_w = c_1/(\omega_h w_0); \qquad \dot{D}_{\dot{w}} = -a_2/(\omega_h^2 w_0) \qquad (3.20)$$
$$\dot{L}_{\dot{w}} = -b_2/(\omega_h^2 w_0); \qquad \dot{M}_{\dot{w}} = -c_2/(\omega_h^2 u_0)$$

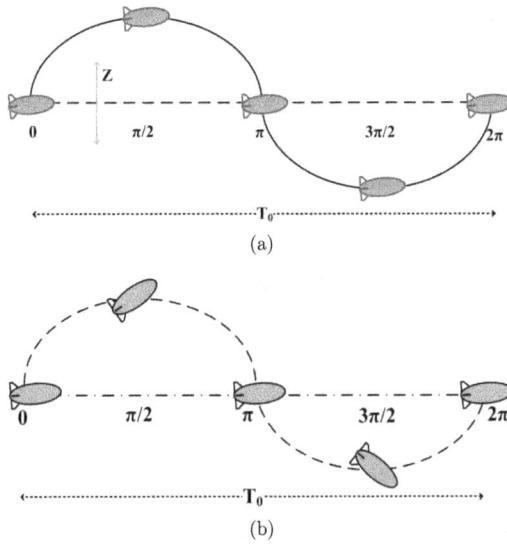

Figure 3.12: (a) Aerostat heave motion and (b) pitching motion.

According to Equation (3.20), a full-cycle response of drag, lift and moment for the forced oscillation along the vertical direction will give the six stability derivatives due to vertical velocity and acceleration.

3.5.3 Pure pitch oscillations

The stability derivatives associated with the rotational pitch motion (about the Y-axis) can be expressed as the linearly decomposed forces and moment as follows,

$$D_a^p = D_e + \dot{D}_\alpha \alpha + \dot{D}_{\dot{\alpha}} \dot{\alpha} + \dot{D}_q q + \dot{D}_{\dot{q}} \dot{q}$$
$$L_a^p = L_e + \dot{L}_\alpha \alpha + \dot{L}_{\dot{\alpha}} \dot{\alpha} + \dot{L}_q q + \dot{L}_{\dot{q}} \dot{q} \qquad (3.21)$$
$$M_a^p = M_e + \dot{M}_\alpha \alpha + \dot{M}_{\dot{\alpha}} \dot{\alpha} + \dot{M}_q q + \dot{M}_{\dot{q}} \dot{q}$$

A forced sinusoidal oscillatory motion will be induced in the aerostat pitch axis, as shown in Figure 3.12 (b). The motion can be defined by the displacement $\alpha = \nu = q_0 \sin \omega_p t$, velocity $\dot{\alpha} = q = \omega_p q_0 \cos \omega_p t$, and acceleration $\dot{q} = -\omega_p^2 q_0 \sin \omega_p t$. Substituting the values of w and $\dot{w}$

into Equation (3.21) will result in,

$$D_a^p = D_e + \dot{D}_\alpha q_0 \sin \omega_p t + \dot{D}_{\dot{\alpha}} \omega_p q_0 \cos \omega_p t + \dot{D}_q \omega_p q_0 \cos \omega_p t - \dot{D}_{\dot{q}} \omega_p^2 q_0 \sin \omega_p t$$

$$L_a^p = L_e + \dot{L}_\alpha q_0 \sin \omega_p t + \dot{L}_{\dot{\alpha}} \omega_p q_0 \cos \omega_p t + \dot{L}_q \omega_p q_0 \cos \omega_p t - \dot{L}_{\dot{q}} \omega_p^2 q_0 \sin \omega_p t \qquad (3.22)$$

$$M_a^p = M_e + \dot{M}_\alpha q_0 \sin \omega_p t + \dot{M}_{\dot{\alpha}} \omega_p q_0 \cos \omega_p t + \dot{M}_q \omega_p q_0 \cos \omega_p t - \dot{M}_{\dot{q}} \omega_p^2 q_0 \sin \omega_p t$$

By comparing the coefficients of in-phase and out-of-phase components of Equation (3.16) and Equation (3.22),

$$\dot{D}_\alpha q_0 - \dot{D}_{\dot{q}} \omega_p^2 = a_2; \quad \dot{D}_{\dot{\alpha}} + \dot{D}_q = a_1/(\omega_p q_0)$$

$$\dot{L}_\alpha q_0 - \dot{L}_{\dot{q}} \omega_p^2 = b_2; \quad \dot{L}_{\dot{\alpha}} + \dot{L}_q = b_1/(\omega_p q_0) \qquad (3.23)$$

$$\dot{M}_\alpha q_0 - \dot{M}_{\dot{q}} \omega_p^2 = c_2; \quad \dot{M}_{\dot{\alpha}} + \dot{M}_q = c_1/(\omega_p q_0)$$

By considering two pitch oscillating frequencies ω_{p1} and ω_{p2}, the stability derivatives due to pitching accelerations can be calculated as,

$$\dot{D}_{\dot{q}} = \frac{a_2^{\omega_{p2}} - a_2^{\omega_{p1}}}{q_0(\omega_{p1}^2 - \omega_{p2}^2)}$$

$$\dot{L}_{\dot{q}} = \frac{b_2^{\omega_{p2}} - b_2^{\omega_{p1}}}{q_0(\omega_{p1}^2 - \omega_{p2}^2)} \qquad (3.24)$$

$$\dot{M}_{\dot{q}} = \frac{c_2^{\omega_{p2}} - c_2^{\omega_{p1}}}{q_0(\omega_{p1}^2 - \omega_{p2}^2)}$$

According to Equations (3.23) and (3.24), a full-cycle response of drag, lift and moment for the forced oscillation about the pitch axis will give the six stability derivatives due to pitching velocity and acceleration.

3.5.4 Simulation of the unsteady condition of the aerostat

The unsteady simulation analysis of the aerostat was done by defining the unsteady wind and unsteady motion of the aerostat simultaneously. A sinusoidal wind gust was used to simulate the unsteady wind. A sinusoidal oscillatory motion was used to induce unsteady motion in the aerostat model. In the current research, these two sinusoidal signals (wind sine and motion sine) were considered input signals. Since both input signals are sinusoidal, periodic, and continuous, there is a chance for both signals to be in-phase with each other or out-of-phase. To investigate the effect of this possible phase shift between the two sinusoidal

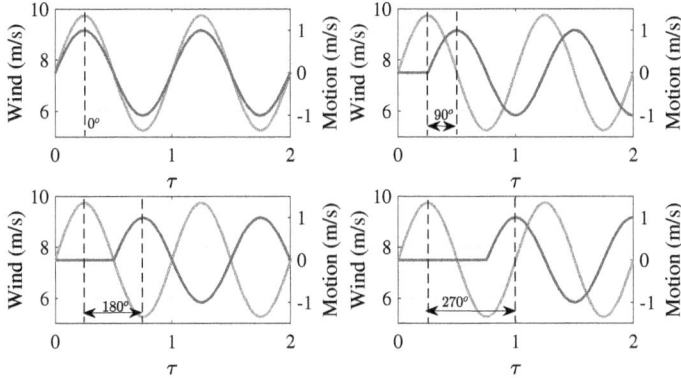

Figure 3.13: Unsteady simulation condition for the aerostat.

inputs, four test conditions were considered: a) both inputs are in phase (0 degree phase shift), b) wind input is leading the motion input by 90 degrees (90 degree phase shift), c) wind input is leading the motion input by 180 degrees (180 degree phase shift), and d) wind input is leading the motion input by 270 degrees (270 degree phase shift). These four test conditions are shown in Figure 3.13. A sinusoidal wind gust of amplitude 30 % of the steady velocity of 7.5 m/s and the heave oscillation motion with a sinusoidal input of 1 m/s amplitude are shown in the figure. The mean wind velocity and the wind gust amplitudes used in this research are selected based on the NIWE wind data collected at an altitude of 100 m [96]. UDFs were used to represent the sinusoidal wind and motion inputs during the CFD simulation analysis.

3.6 Validation of the simulation setup

3.6.1 Aerostat geometry and computational setup

Validation of the aerostat 3D model used for the simulation study was done by comparing it with the steady results available in the literature [33], as shown in Figure 3.14. A gondola was added to the model for the validation study since the experimental data is for a model with the gondola. Normalised drag and lift forces show appreciable matching with the experimental data for the AoA under consideration, as shown in Figure 3.14. The drag and lift variation with respect to the AoA is usually symmetric, as seen in the figure. The maximum deviation of C_D is 3.89 %, and C_L is 5 % of the experimental values.

To assess the quality of the developed computational grid for the aerostat model used for the aerodynamic analysis, a grid independence study was conducted for different grid configurations. The test conditions were as follows: standard EOG with 12.5 m/s and 10.5 seconds duration, aerostat at 10 degrees AoA, and a time step of 0.01 seconds. The value of the drag force was compared for different meshes. The normalised value for the drag is presented here. Figure 3.15 (a) shows the grid independence study conducted among a fine grid of 6.8 million grid points, a medium grid of 4.1 million grid points, and a coarse grid of 2.7 million grid points. Since a reasonable closeness in the results was able to be generated by the three grid configurations, it can be concluded that increasing the grid density is not justified. Therefore, the medium grid was selected for this research.

A time-independence study was conducted to justify the computational time step used for the simulation. The test conditions were as follows: standard EOG with 12.5 m/s and 10.5 seconds duration, aerostat at 10 degrees AoA and a medium grid of 4.1 million. A time step independence test on the selected grid for two different time steps, as shown in Figure 3.15 (b), was conducted. Drag is measured in both cases. Both time steps showed considerable similarity in the results, and the use of a smaller time step is not justified.

The efficacy of the computational grid used for the stability derivative analysis was tested for different grid refinements, as shown in Figure 3.15 (c), and the simulation time step independence study was performed, as shown in Figure 3.15 (d). The simulation was done by providing the aerostat with an unsteady velocity in the form of a sinusoidal gust. The coeffi-

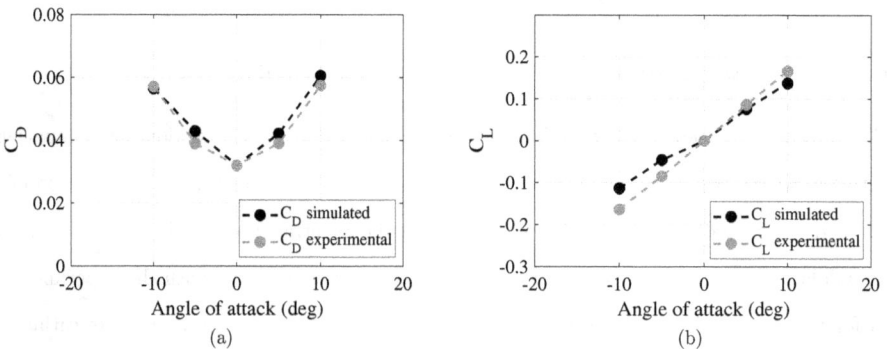

Figure 3.14: Validation of the aerostat model: (a) drag and (b) lift.

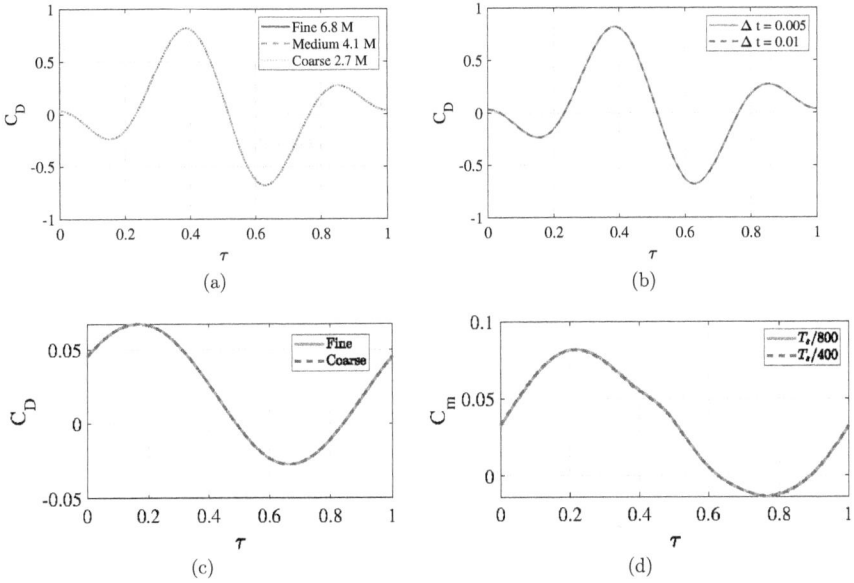

Figure 3.15: (a) Grid and (b) time independence study results using IEC gust. (c) Grid and (d) time independence study results using sine gust.

cients of drag and moment were measured, respectively for the comparison. The results show that the grid used for the research is independent of the grid refinement and the time step.

A comparison among the realizable $k - \epsilon$, RNG $k - \epsilon$, Shear-Stress Transport (SST) $k - \omega$, and generalized $k - \omega$ turbulence models, as explained in Section 3.4.1, was conducted, as shown in Figure 3.16.

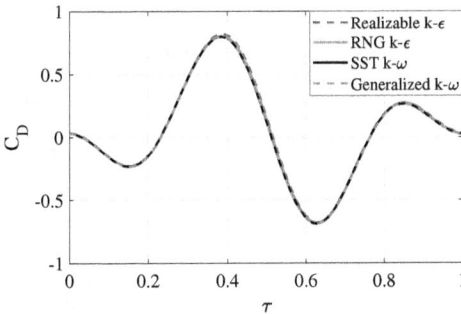

Figure 3.16: Turbulence model comparison.

3.6.2 Grid motion

The dynamic mesh method for the heave and surge oscillations and the grid motion feature for the pitch oscillations were validated using the AGARD CT1 test data set [99]. NACA 0012 airfoil was used for the validation. A comparison of the lift coefficient and the pitching moment coefficient with the experimental data is shown in Figure 3.17. An appreciable similarity is observed in the results, thus validating the grid motion feasibility for further analysis.

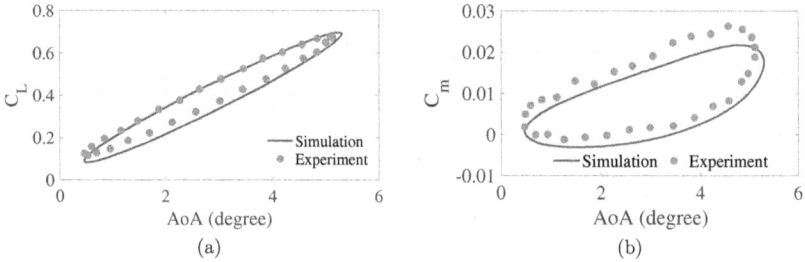

Figure 3.17: Validation of grid motion: (a) lift and (b) moment.

3.6.3 Stability derivative extraction methodology

The methodology for the extraction of stability derivatives presented in Chapter 3 was validated using the 6:1 prolate spheroid. The added mass coefficients were compared with the theoretical calculation as well as with the simulation study by Wang [78].

For the estimation of $\dot{L}_{\dot{w}}$, a sinusoidal oscillation with an amplitude 1 m/s and a frequency of 3.9 rad was used. For the estimation of $\dot{M}_{\dot{q}}$, two sinusoidal oscillations with an amplitude of 5 deg and frequencies $\omega_1 = 29.79$ rad and $\omega_2 = 21.06$ rad were used. The pitching moment response obtained for the two oscillatory inputs is shown in Figure 3.18.

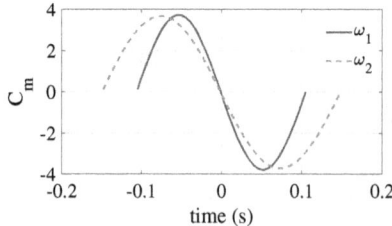

Figure 3.18: Prolate spheroid pitching moment.

The theoretical calculations can be obtained using the following equations [100],

$$\dot{L}_{\dot{w}} = -\frac{4}{3}\pi\rho abc \frac{C_0}{2 - C_0} \tag{3.25}$$

$$\dot{M}_{\dot{q}} = -\frac{4}{15}\pi\rho \frac{abc(a^2 - c^2)^2(A_0 - C_0)}{2(c^2 - a^2) + (C_0 - A_0)(c^2 + a^2)} \tag{3.26}$$

where a, b, c are the dimensions of the ellipsoid, and A_0 and C_0 can be given as follows,

$$A_0 = \frac{2\nu}{(\nu^2 - 1)^{3/2}} \left[\ln(\sqrt{\nu^2 - 1} + \nu) - \frac{\sqrt{\nu^2 - 1}}{\nu} \right] \tag{3.27}$$

$$C_0 = \frac{\nu^2}{\nu^2 - 1} \left[1 - \frac{\nu^2}{\sqrt{\nu^2 - 1}} \ln(\sqrt{\nu^2 - 1} + \nu) \right] \tag{3.28}$$

$$\tag{3.29}$$

The simulated value of $\dot{L}_{\dot{w}}$ is -0.0457 and the theoretical value is -0.0430, and for $\dot{M}_{\dot{q}}$, the corresponding values are -0.00339 and -0.0033. A comparison of the simulated values of the stability derivatives with the theoretical values is shown in Table 3.4. The results are in good agreement with the theoretical values and the reference literature.

Table 3.4: Prolate spheroid results comparison.

Stability derivative	Simulated value	Theoretical value	Wang [78]
$\dot{L}_{\dot{w}}$	-0.0457	-0.0430	-0.042
$\dot{M}_{\dot{q}}$	-0.00339	-0.0033	-0.00332

3.6.4 Acceleration and frequency independence study

As the analysis involves the forced oscillation of the aerostat along the three DoF, the independence study of the stability derivatives from the amplitude of oscillations (acceleration of the motion) has to be done. Such a study was conducted for the Zhiyuan aerostat for surge, heave and pitch oscillations, and the acceleration independence was tested for all the stability derivatives involved. As a representative result, the acceleration independence study conducted for surge motion is shown in Figure 3.19 (a) and (b). Figure 3.19 (a) shows the full cycle drag response, and Figure 3.19 (b) shows the drag derivative. The derivative, $\dot{D}_{\dot{u}}$, with almost the same values, which shows the simulation's acceleration independence.

The estimation of moment stability derivatives involves the application of two sinusoidal

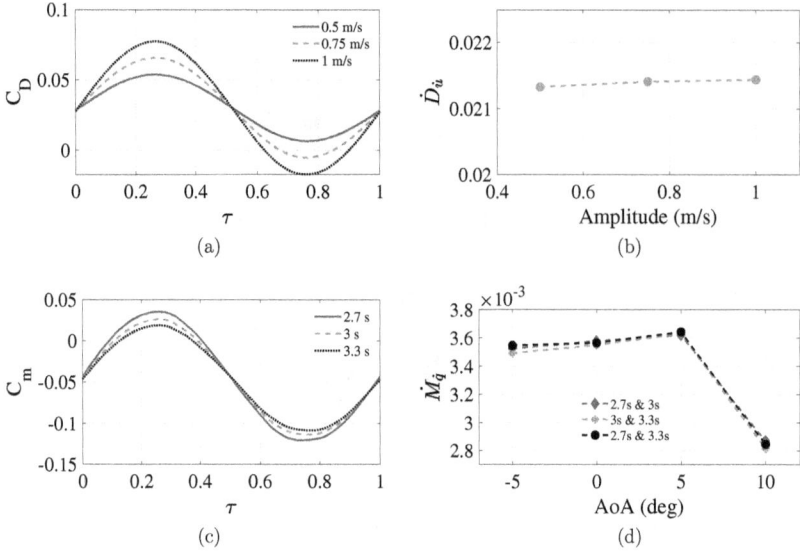

Figure 3.19: Acceleration and frequency independence study results.

oscillations with different frequencies. A frequency independence study was conducted for all the stability derivatives that required two frequencies. The results for the estimation of $\dot{M}_{\dot{q}}$ for a 10-degree AoA are shown in Figure 3.19 (c) and (d). Figure 3.19 (c) shows the full cycle moment response, and Figure 3.19 (d) shows the moment derivative. Here, three frequencies are used for the comparison: 2.32 rad/s (2.7 s), 2.09 rad/s (3 s) and 1.90 rad/s (3.3 s). The frequency independence of the simulation is clearly observable in the figure.

3.7   Summary

This chapter presented the various methodologies and models involved in the current research. The aerostat geometries used for the analysis are presented with their geometrical details. Four aerostat models are presented, namely, Zhiyuan, GNVR, NPL, and HAA. The wind gust models used in the analysis are presented along with their velocity profiles. It includes discrete gust models and continuous gust models. The numerical simulation setup used for the CFD-based analysis is also explained in this chapter. A detailed description of the stability derivative extraction methodology used in the analysis is given. Finally, a set of validation results supporting the above-mentioned methodologies is also presented.

Chapter 4

Aerodynamic Performance Analysis of the Aerostat

4.1 Introduction

As the aerostat is a buoyant aerodynamic structure, it is highly prone to oscillations due to wind fluctuations. Most of the aerial applications of the aerostat are operated at an altitude of hundreds of meters above the ground. At these altitudes of the atmosphere, the occurrence of wind gusts is significant [101]. Hence, the effect of wind gusts cannot be avoided while designing any systems that are supposed to be deployed at these altitudes [102].

So, an analysis of the effect of wind gusts on the characteristics of aerostats is of crucial importance. The design of tethers connecting the aerostat to the ground, the aerostat's payload carrying capacity, the design of the wings, the optimisation of shape, and the control system design are done based on the study of the effect of wind on the system. There are different methodologies available in the open literature for analysing the impact of wind on aerostats, such as flight tests, wind tunnel or tow tank tests, and numerical simulations.

There is considerable literature regarding the investigation of the aerodynamic character- istics of aerostats under uniform wind flow conditions. There are no significant investigations presented on the effect of wind gusts on the aerodynamic characteristics of the aerostats. The aim of this chapter is to investigate the effect of wind gusts on the aerodynamic characteristics of an aerostat using CFD. The major contribution of this chapter is the aerodynamic perfor- mance evaluation of an aerostat considering different AoA, different gust amplitudes, different

49

gust shapes, and different gust lengths. The work in this chapter is an initial study of a tethered aerostat-based system. It is assumed that the wind is confined to the axial direction of the aerostat. For streamline shapes such as aerostats and airships, the range of AoA for the hovering or station keeping phase was reported to be 0 to 10 degrees [37, 46, 67, 103]. There are conditions with a negative AoA, such as the pitch control, take-down stage for maintenance, refilling of the LTA gas, and relocating the system [61]. In this chapter, negative and positive AoA such as −5, 0, 5 and 10 degrees are considered. This ensures that all the required flight scenarios are included in the analysis of the aerostat. A set of gust parameters such as amplitude, shape, and duration are selected based on the statistical and meteorological data available for the proposed sites. Wind data from the Gujarat coastal area provided by the NIWE is used for that purpose [96].

4.2 Results obtained from the analysis

The simulation and analysis of the effect of AoA, shape of the gust profile, amplitude of the gust, and duration of the gust on the aerodynamic characteristics of the aerostat are presented in this section. The drag, lift, and moment about the Y-axis are used for the analysis, and the normalised expressions for these forces and moment are given as follows.

$$C_D = \frac{D}{0.5\rho V^2 A}, \quad C_L = \frac{L}{0.5\rho V^2 A}, \quad C_m = \frac{M}{0.5\rho V^2 A L_A}$$

where D, L and M are the drag, lift and moment, respectively, A is the reference area, and L_A is the reference length.

The simulation details for each case considered in this chapter are consolidated in Table 4.1 for better understanding.

The results obtained are classified into five sections based on the parameter considered for the analysis, and the last section presents the scaling methodology of the responses.

4.2.1 Effect of AoA

As specified in Section 4.1, −5, 0, 5, and 10 degree AoA were considered for this analysis, which represents most of the flight scenarios of the aerostat. Figure 4.1 shows the response of

Table 4.1: Aerodynamic analysis: Consolidated simulation details.

	Aerostat shape	Gust velocity (m/s)	Gust duration (s)	Angle of attack (deg)
Model validation	Zhiyuan	12.5	10.5	10
Grid independence	Zhiyuan	12.5	10.5	10
Time independence	Zhiyuan	12.5	10.5	10
Effect of AoA	Zhiyuan	12.5	10.5	-5, 0 5, 10
Effect of amplitude	Zhiyuan	6, 8, 10, 12.5, 15	10.5	10
Effect of time period	Zhiyuan	12.5	6, 8, 10.5, 12, 15	10
Effect of gust shape	Zhiyuan	12.5	10.5	10
Effect of aerostat shape	ZBG	6, 8, 10, 12.5, 15	10.5	10
	ZMG1			10
	ZMG2			10

the aerostat for the IEC gust of amplitude of 12.5 m/s and gust duration of 10.5 seconds.

Figure 4.1 (a) represents the variation of drag for different AoA. The X-axis represents the normalised time specified in Equation (3.2). The mean value of the drag for each case is 0.035, 0.028, 0.035, and 0.048, respectively. Figure 4.1 (b) shows a comparison between the IEC gust and the drag response. A 12.5 m/s and 10.5 second IEC gust is superimposed on the corresponding drag curve to visualize the shift in the response. As the velocity drops with a constant deceleration towards its first negative peak B, the pressure drag reduces drastically, and the viscous drag reduces slightly, resulting in a drop in the total drag to a local minimum of A, as shown in Figure 4.1 (b). The pressure drag contribution is -0.2395, and the viscous drag is 0.00515 at instant A, and that at instant B is 0.00981 and 0.00288, respectively, as shown in Figure 4.2 (a) for 10 degree AoA. An increase in the pressure drag from point A to point B is evident, which causes a rise in the drag after point A. Figure 4.3 (a) shows the velocity distribution in the flow field of the aerostat, and Figure 4.3 (b) shows the pressure coefficient on the aerostat surface at instant A. The pressure distribution on the front side is much lesser than that on the rear side, and the pressure gradient between the upper and lower surfaces is considerably less. Thus, the contribution to the drag due to the pressure effect is

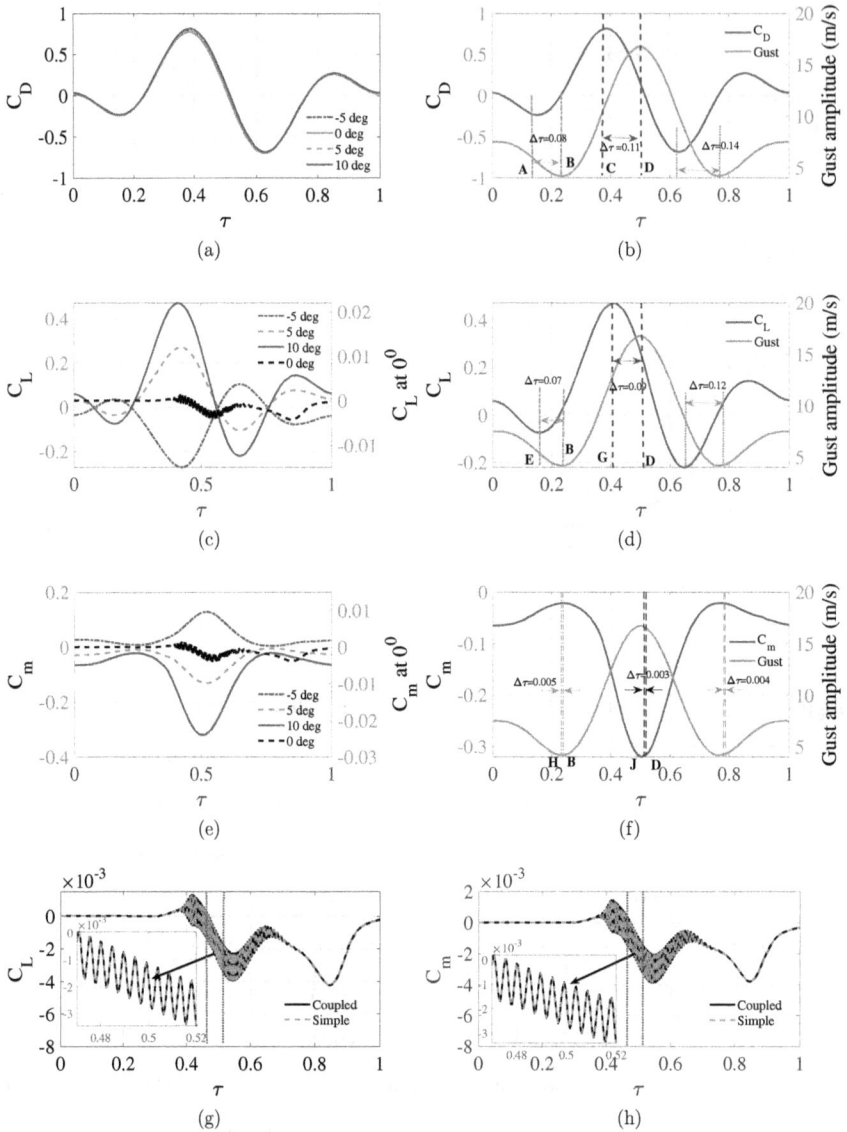

Figure 4.1: Aerodynamic analysis: Effect of AoA: (a) coefficient of drag, (b) drag and gust, (c) coefficient of lift, (d) lift and gust, (e) coefficient of moment, (f) moment and gust, and coefficient of (g) lift and (h) moment at 0 degree AoA.

less than that of the viscous effect. As the velocity reduces further, the viscous drag continues to decrease, and the pressure drag rises slightly until the velocity reduces to a minimum at B. The velocity distribution and the pressure contour at instants E and B are shown in Figure

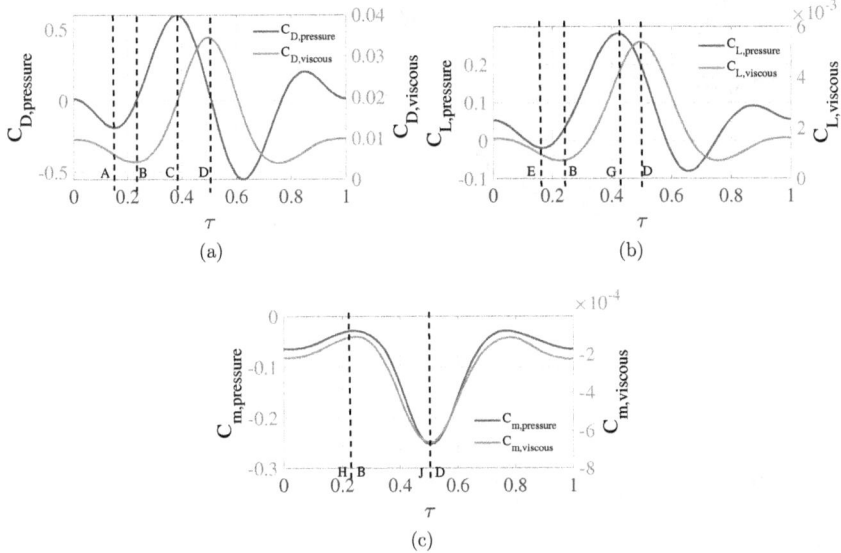

Figure 4.2: Pressure and viscous components of the forces and moment: coefficient of (a) drag, (b) lift, and (c) moment.

4.3 (c) - (f). It is seen that the pressure gradient between the upper and lower surfaces of the aerostat is slightly increasing. Still, the viscous drag is dominating. Beyond this point, the gust velocity rises with a constant acceleration towards point D, and the pressure drag shoots up to increase the total drag linearly towards its maximum at C. Figure 4.4 shows the velocity distribution and pressure profiles at instants C, G and D. It is seen that the pressure at the front side is higher than that at the rear side of the aerostat and the pressure drag dominates over the viscous drag and causing the drag to shoot up to its maximum value at C. Then the pressure drag falls slightly, and the total drag started to decrease. During this transition, even though the viscous drag is aiding the drag, its contribution is minimal. The pressure drag is 0.7969, and the viscous drag is 0.02259 at instant C and that at D is 0.11142 and 0.04328 and at instant G is 0.7609 and 0.0285 respectively, as shown in Figure 4.2 (a). As the velocity drops after D, the pressure drag reduces drastically, causing a linear drop in the drag. This peculiar behaviour in the drag response causes a temporal shift in the drag response with respect to the gust profile.

Figure 4.1 (c) shows the lift response for different AoA. The average value of lift for each case is -0.054, -0.001, 0.053 and 0.086, respectively. A considerable variation of the lift is

53

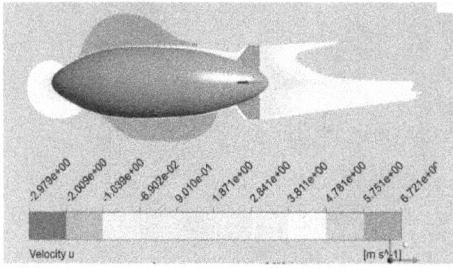
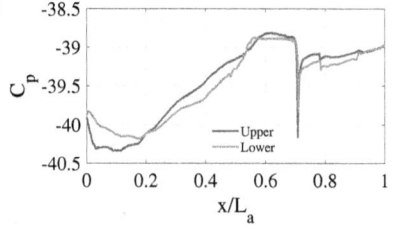

(a)   (b)

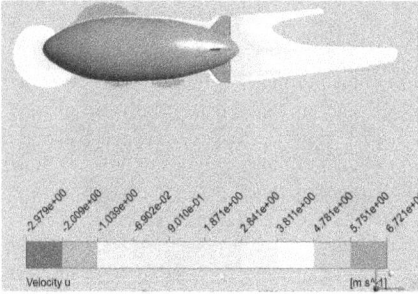
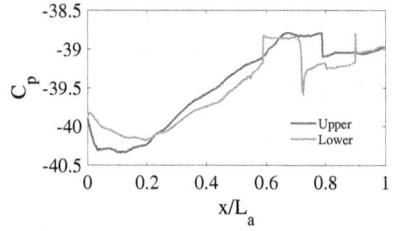

(c)   (d)

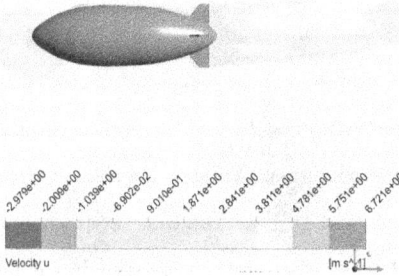
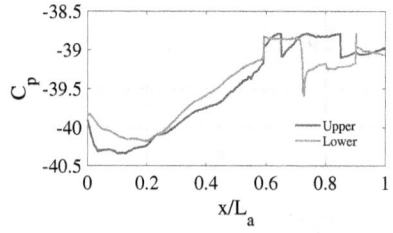

(e)   (f)

Figure 4.3: Velocity contour for 10 degree AoA at instant (a) A, (c) E and (e) B; Pressure profile at instant (b) A, (d) E and (f) B.

visible than that for the drag. Lift is increasing with the AoA, as shown. This is mainly due to the development of a pressure gradient between the top and bottom surfaces of the aerostat for different AoA. For the negative AoA, the value of lift is negative because of the negative pressure gradient; i.e., high pressure is formed on the top surface than on the bottom surface of the aerostat. Also, the negative AoA forces the centre of pressure to shift towards the leading edge and to contribute a moment in the opposite direction than that for a positive AoA. Figure 4.1 (d) shows the lift response for a 12.5 m/s IEC gust for a 10 degree AoA

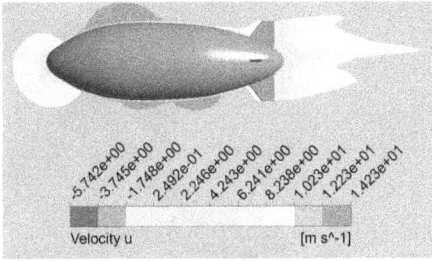
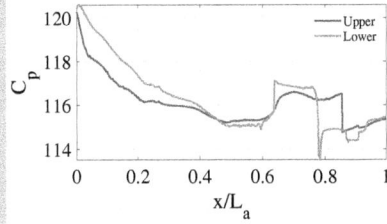

(a)

(b)

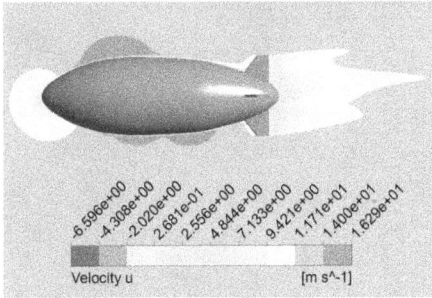
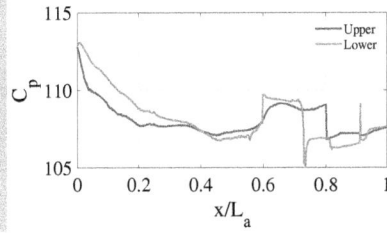

(c)

(d)

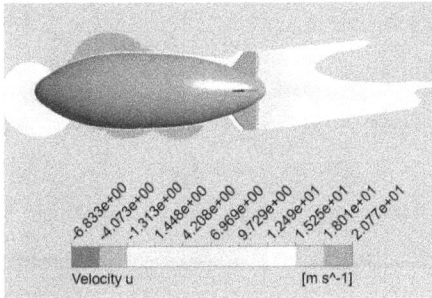
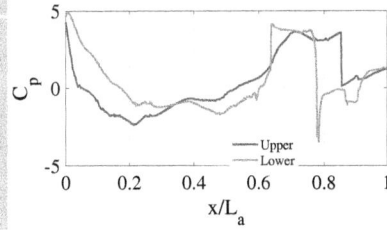

(e)

(f)

Figure 4.4: Velocity contour for 10 degree AoA at instant (a) C, (c) G and (e) D; Pressure profile at instant (b) C, (d) G and (f) D.

case. As the gust velocity drops from its mean value towards the minimum at B, the pressure gradient between the upper and lower surfaces of the aerostat increase, causing the dropping of lift to its negative local minimum at E (refer to Figure 4.2 (b)). As the velocity reduces further and around halfway, the pressure distribution at the lower surface starts to grow, and as a result, the lift rises slowly. The pressure profile around the top and bottom surfaces of the aerostat is shown in Figure 4.3 (d). The pressure gradient formation is shown in the figure. Beyond point B, the gust velocity rises in a linear manner aiding the pressure gradient and,

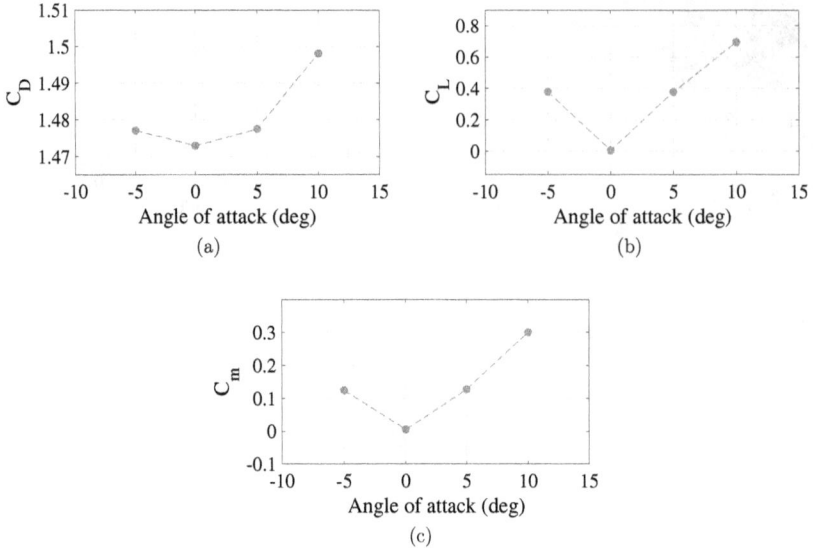

Figure 4.5: Peak to peak values of (a) drag, (b) lift and (c) moment for different AoA.

thereby, the lift generation. Thus, the lift rises to its maximum at G. The pressure profile at the instant G is as shown in Figure 4.4 (d). Further increase in the gust velocity causes the separation of flow on the upper surfaces, causing a fall in the lift value, which continues as the velocity drops towards its minimum. The drop in the pressure is clearly seen in Figure 4.4 (f).

Figure 4.1 (e) represents the moment response for different AoA. The mean value of the moment for each AoA is 0.0359, $-8.69e^{-04}$, -0.0364 and -0.0908, respectively. Similar to that for the lift response, there is a considerable variation in the moment response for different AoA. Figure 4.1 (f) compares the occurrence of peaks for the gust and the corresponding moment curve at 10 degree AoA. As the gust velocity drops towards its first minimum at B, the negative lift force causes a downward moment, as shown at H (refer to Figure 4.2 (c)). As the velocity increases after B, the increasing positive lift aids the upward moment. The response for the zero AoA case shows oscillations during the peak of the gust. These oscillations are induced due to the vortex around the tail wings of the aerostat. Figure 4.1 (g) and (h) show the lift and moment response at zero degree AoA. The presence of oscillations was verified using the SIMPLE algorithm and coupled algorithm of ANSYS Fluent [98], as shown in the figure.

Figure 4.5 (a) represents the peak to peak value of the drag curve obtained for each AoA case. Minimum drag is obtained at zero AoA since the main contribution of drag, in that case, is from the skin friction drag. The drag of the aerostat increases with the increase in the AoA, which is due to the transition of the flow separation point towards the leading edge of the aerostat, the increase in the frontal area and the thickening of the boundary layer. The rate at which the drag varies increases with the AoA. The symmetric nature of the drag versus AoA curve for streamline shapes, as reported in [33, 51], has been obtained in the current research. Figures 4.5 (b) and (c) show the peak to peak value for the lift and moment curves for different AoA. A considerable variation is visible for the lift and moment than for the drag.

The first column of Figure 4.6 shows the evolution of the amplitude of the peaks in the force and moment responses, and the second column shows the temporal shift for each response curve with respect to the corresponding gust velocity curve. Three peaks (positive as well as negative) are identified for the curves. $\Delta\tau$ represents the normalised time difference between the response peak and the gust peak. As the gust progresses, the $\Delta\tau$ value increases for the drag and lift curves. For the moment curve, the $\Delta\tau$ variation is almost constant.

It can be concluded from the analysis of different AoA that the drag is insensitive to the variations in the AoA. It is mainly due to the fact that the drag is contributed majorly by the flow over the surface rather than the pressure distribution around the aerostat surface. On the other hand, the lift is highly dependent on the AoA because, here, the primary role is played by the pressure distribution. Positive AoA provides the maximum lift with an almost constant drag value. It substantiates the application of positive operating AoA for aerostats for better performance. Among the two positive angles considered here, the 5 degree AoA shows comparatively fewer oscillations in the gust than the 10 degree case. There is no hint of the stalling of the aerostat, and it may occur for some larger AoA, which is not under the scope of this work.

4.2.2   Effect of gust amplitude

The effect of the amplitude of the gust on the aerostat was investigated using the IEC gust given by Equation (3.1). Five different gusts of amplitude 6 m/s, 8 m/s, 10 m/s, 12.5 m/s and 15 m/s were considered for the analysis. The time duration of the gust was kept fixed at 10.5 seconds. The amplitude of the gust is selected based on the wind data available, and the

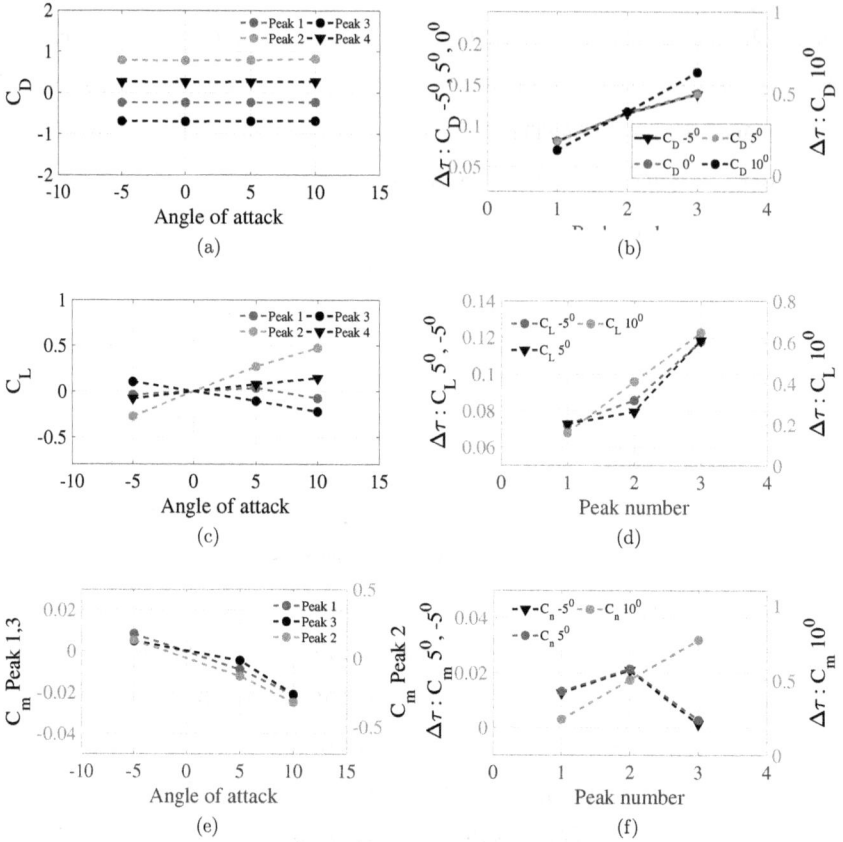

Figure 4.6: Peak amplitudes and time shifts of responses for different AoA: amplitude of each peaks of (a) drag, (c) lift and (e) moment curve; temporal shift in the peaks of (b) drag, (d) lift and (f) moment with respect to the gust.

aerostat is considered to be at 10 degree AoA. Since there is no change in the AoA, the amplitude variation of the gusts was majorly causing a variation in the pressure distribution around the aerostat surface. Figure 4.7 shows the simulation results for different gust amplitudes.

Figure 4.7 (a) represents the variation of drag with the amplitude of the gust. The average value for the drag is 0.038, 0.04, 0.042, 0.047 and 0.05, respectively. Drag is increasing with the amplitude as expected since the drag is proportional to the square of the velocity. As the gust velocity decreases to its first minimum, the pressure drag reduces, causing the drag to drop to its first minimum. The linear increase in the gust velocity then causes the pressure drag to increase linearly with the gust velocity. Figure 4.7 (b) shows the variation in the peak values

58

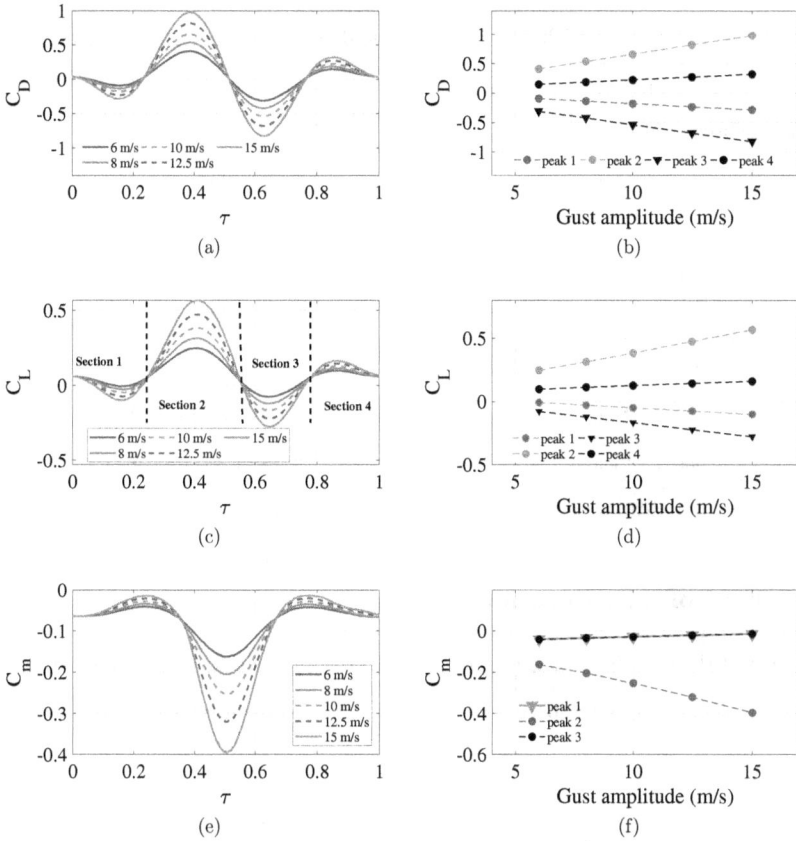

Figure 4.7: Aerodynamic analysis: Effect of gust amplitude; time series plot for (a) drag, (c) lift, (e) moment responses; variation of peak values of (b) drag, (d) lift and (f) moment responses.

of the drag curves for different gust amplitudes. Four peaks are identified (both positive and negative) for the drag response. A linear increase in the peak value is shown in the figure.

Figure 4.7 (c) shows the lift curve for the different gusts. The average lift value for each gust is 0.066, 0.07, 0.075, 0.083 and 0.092, respectively. The positive AoA condition aids the gust to increase the lift with increasing gust amplitude. As the gust amplitude increases, the pressure gradient between the upper and lower surfaces also increases. This is the main cause of the increase in the lift. Figure 4.7 (d) shows the variation in the peak values of the lift curves for different gust amplitudes. Four peaks are identified (both positive and negative) for the lift response. A linear increase in the peak value is shown in the figure.

Figure 4.7 (e) shows the moment response for different gust amplitudes. The average value for the moment is −0.074, −0.079, −0.084, −0.091 and −0.099 for each gust case, respectively. Figure 4.7 (f) shows the variation in the peak values of the moment curves for different gust amplitudes. Three peaks are identified for the lift response. The peak that almost coincides with the gust peak shows a considerable increase with the increase in gust amplitude.

Figure 4.8 shows the peak to peak values of the drag, lift and moment curves for the five different gusts. There is a linear increase in the amplitude of the curves as the gust amplitude increases. The drag response is highly influenced by the variation in the gust amplitude, and the moment response is least affected.

It can be concluded that the variation of the gust amplitude affects the drag, lift and moment characteristics of the aerostat largely. The amplitudes of the response of the aerostat for different gust amplitudes vary linearly with the gust value.

4.2.3   Effect of gust duration

The effect of the duration of the gust on the aerodynamic performance of the aerostat was investigated using the IEC gust specified in Equation (3.1). Five different gusts with a constant

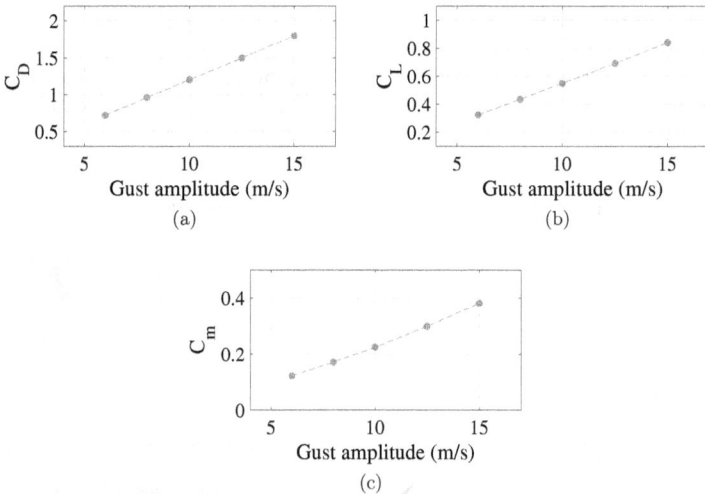

Figure 4.8: Peak to peak values of responses for different gust amplitudes: coefficient of (a) drag, (b) lift, and (c) moment.

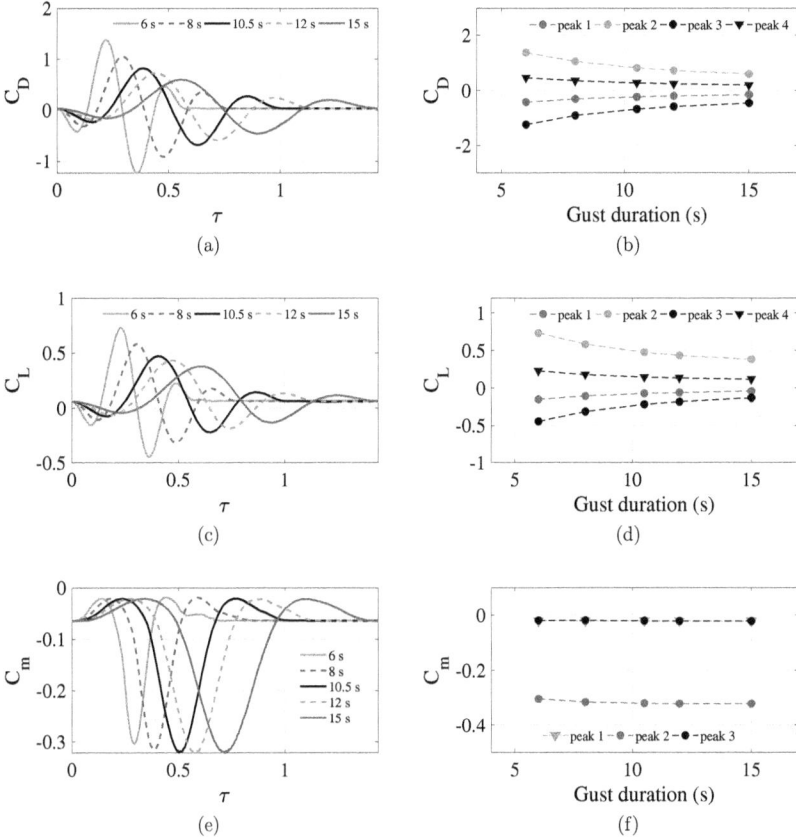

Figure 4.9: Aerodynamic analysis: Effect of gust duration; time series plot for (a) drag, (c) lift, (e) moment; variation of the peak values of (b) drag, (d) lift and (e) moment response curve with respect to gust duration.

amplitude of 12.5 m/s were considered with time duration of 6, 8, 10.5, 12 and 15 seconds, respectively. The aerostat is considered to be at 10 degree AoA.

Figure 4.9 shows the simulation results for the considered gust conditions. Figure 4.9 (a) shows drag response for different gust duration. The average value for the drag is 0.0464, 0.0474, 0.0468, 0.0474 and 0.0474, respectively, for each gust case. It is seen that the overshoot in the drag value decreases with the increase in gust duration in spite of the constant gust amplitude, and the longer gust has the minimum variation in the drag. Figure 4.9 (b) shows the evolution of the peak values of drag with increasing gust length. There are four peaks in the drag response. As the gust duration increases, the peak amplitude decreases. As seen in

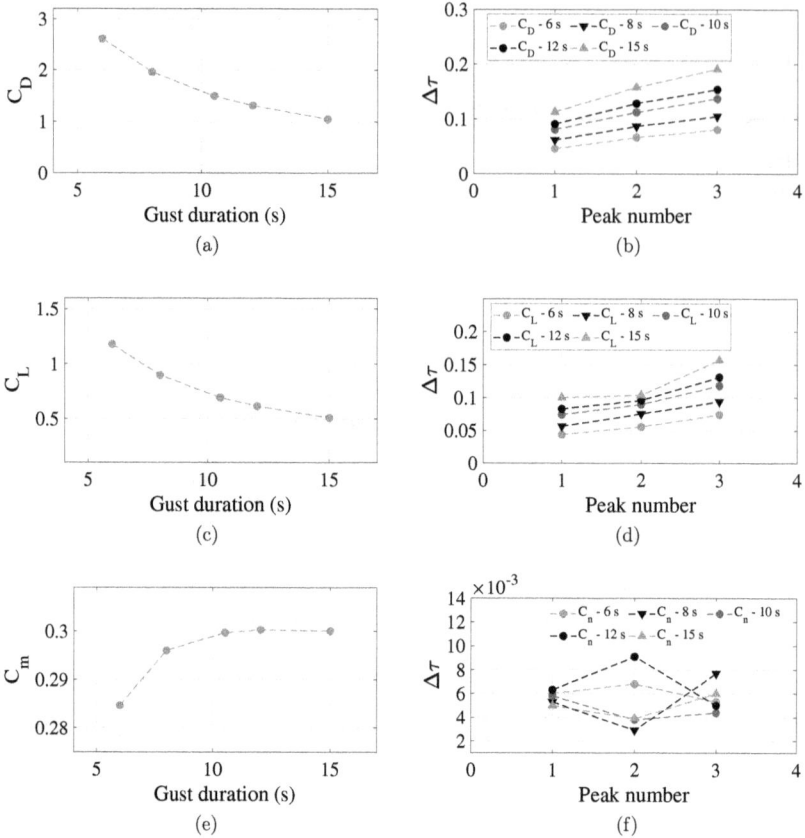

Figure 4.10: Peak amplitudes and time shifts for different gust duration: peak to peak value for (a) drag, (c) lift and (e) moment curves; temporal shift of peaks of (b) drag, (d) lift and (e) moment with respect to the gust peaks.

Figure 4.9 (b), the curve flattens as the gust duration increases. The peak amplitude value may saturate with a further increase in the gust duration.

Figure 4.9 (c) shows the lift response for different gust duration. Similar to the drag response, as the gust length increases, the overshoot in the lift value is reduced. The average value for the lift curves is 0.086, 0.085, 0.0828, 0.0833 and 0.0827, respectively. The variation of the peak values of the lift curve with respect to the gust duration is shown in Figure 4.9 (d).

Figure 4.9 (e) represents the moment response for different gust duration. It should be noted that the moment increases slightly with the increase in the gust length. The average values for the moment curves are -0.0813, -0.0886, -0.0908, -0.0931 and -0.0945, respectively. The variation of the peak values of the moment with the gust duration is shown in Figure 4.9 (f). The variation in the moment is negligible compared to the variation of peaks for drag and lift.

Figure 4.10 (a), (c) and (e) shows the peak to peak amplitude of the drag, lift and moment curves for different gust duration. It is seen that the amplitude decreases with the increase in the gust length for the case of drag and lift. As the gust length increases, a smooth nature will be obtained from the gust, and thus the impact it can make will be considerably reduced. For a very short gust of 6 seconds, the drag has risen to a very large value. Figure 4.10 (b), (d) and (f) show the temporal shift between the corresponding peaks of the gust and the force/moment response.

It can be concluded that the duration of gust or the gust length has a significant influence on the aerodynamic performance of the aerostat. Short gusts are more severe than long gusts with the same amplitude. As the gust length increases, the aerostat response becomes smooth, and the deviation from the mean value diminishes gradually. The reduction in the peak values of the response ceases with a further increase in the gust duration.

4.2.4 Effect of aerostat shape

The effect of wind gusts on the aerodynamic parameters of the aerostat was presented in the previous sections. Zhiyuan geometry is used for those analyses. In order to generalize the presented analysis, it is necessary to consider different shapes for the aerostat. The Zhiyuan Base Geometry (ZBG) can be parameterized using the basic geometrical elements such as parabolas and splines, as shown in Figure 4.11 (a). The governing equations of the corresponding geometries with their endpoint coordinates which are with reference to the leading edge as the origin are shown in the figure. Two new shapes are obtained from the ZBG and labelled as Zhiyuan Modified Geometry (ZMG), as shown in Figure 4.11 (b). The parameters of the aerostat shape geometry elements are given in Table 4.2.

Figure 4.12 shows the simulation results obtained for the ZMG1 aerostat profile. IEC gust with a time duration of 10.5 seconds is used for the simulation with the aerostat at an AoA of 10 degrees. Figure 4.12 (a) shows the drag response for five different amplitudes. The response is similar to that for the ZBG shape (Figure 4.7) but with reduced amplitude in the drag value. This is due to the reduction in skin friction drag because of the new shape. The lift response is

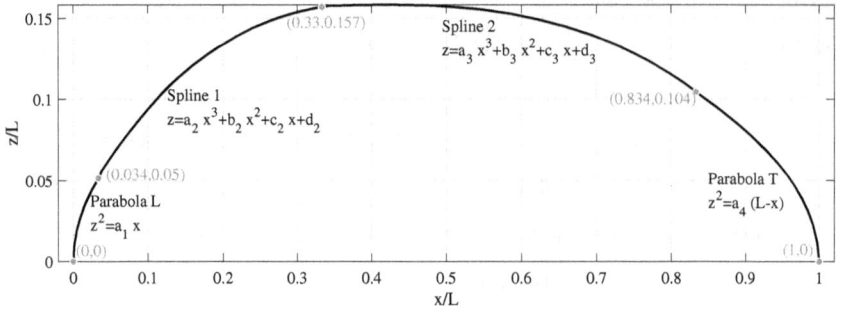

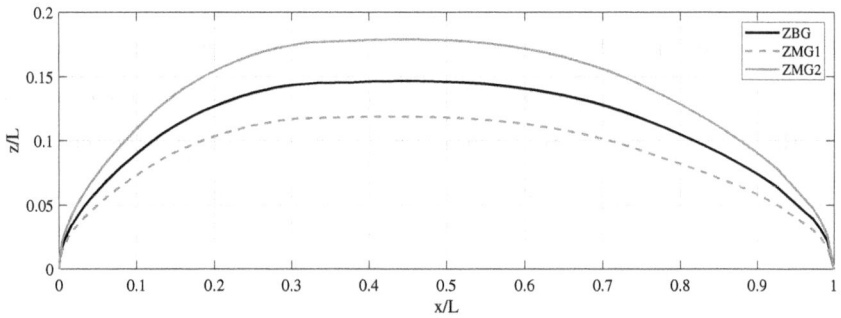

Figure 4.11: (a) Aerostat base shape and (b) the variants considered for the study.

not much affected by the variation in the geometry because the differential pressure formation remains unchanged. The moment response shows a considerable reduction in its amplitude. The variation of the response peaks with respect to the amplitude is shown in Figure 4.12 (b), (d) and (f).

Table 4.2: Geometric parameters for the aerostat shapes.

	ZBG	ZMG1	ZMG2
a_1	977.4	678	1478
a_2	7.4E-09	5.9E-09	8.9E-09
b_2	-1.3E-04	-1.1E-04	-1.6E-04
c_2	0.83	0.67	1.1
d_2	332.83	268.41	401
a_3	-2.8E-09	-2.3E-09	-3.4E-09
b_3	3.4E-05	2.7E-05	4.1E-05
c_3	-0.12	-0.1	-0.15
d_3	2097.54	1691.6	2527.2
a_4	821.3	534.3	1214.3

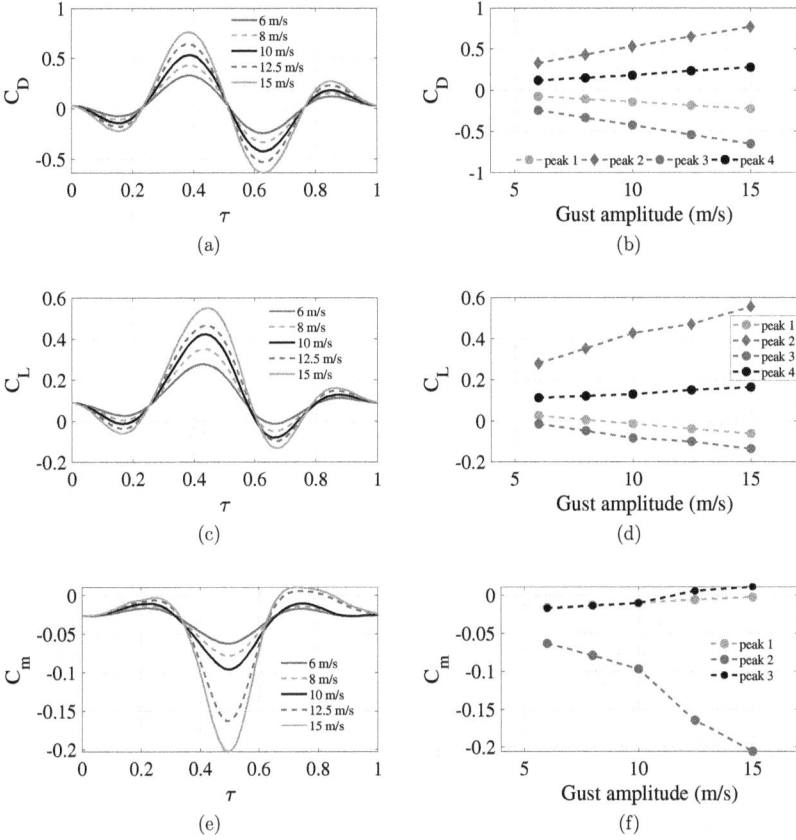

Figure 4.12: Aerodynamic analysis: Response for aerostat ZMG1; (a) drag, (c)lift and (e) moment; peaks of (b) drag, (d) lift and (f) moment curves.

Figure 4.13 shows the simulation results obtained for the ZMG2 aerostat profile. IEC gust with a time duration of 10.5 seconds is used for the simulation with the aerostat at an AoA of 10 degrees. Figure 4.13 (a) shows the drag response for five different amplitudes. The drag response shows similar characteristics as the ZBG shape response. It shows that a further increase in the diameter of the geometry is not aiding the skin friction drag. As in the case of the ZMG1 shape, the lift response for the ZMG2 shape is similar to that of the ZBG response. There is a slight increase in the moment response than from the ZMG1 case. But its amplitudes are less than that for the ZBG case. The peak values of the response curves are shown in Figure 4.13 (b), (d) and (f) for different gust amplitudes.

A comparison of the force and moment responses for the three aerostat geometries for a

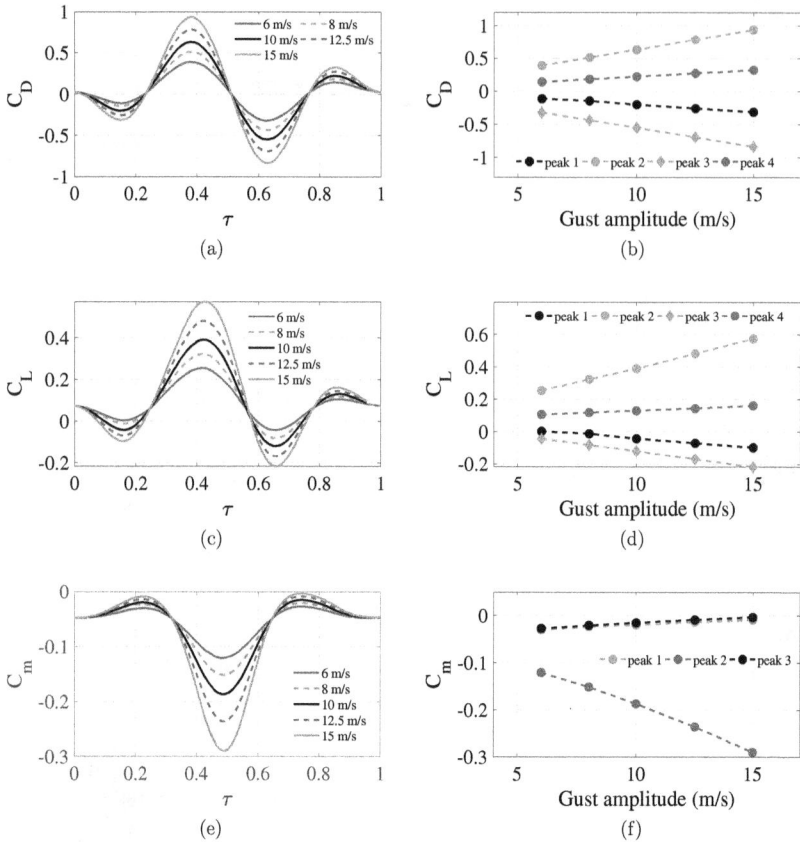

Figure 4.13: Aerodynamic analysis: Response for aerostat ZMG2; (a) drag, (c)lift and (e) moment; peaks of (b) drag, (d) lift and (f) moment curves.

12.5 m/s IEC gust is shown in Figure 4.14 (a), (c) and (e). The mean values of the response curves of three aerostat geometries for different gust amplitudes are shown in Figure 4.14 (b), (d) and (f). The Zhiyuan geometry has the least drag among the three shapes. Its lift characteristics is also desirable. ZMG1 shape has the maximum lift among the three geometries. It also has less drag compared to the ZMG2 shape.

It can be concluded from this analysis that the reduction in the diameter of the geometry, by keeping the length constant, can result in more lift with the expense of a considerable amount of drag. As the diameter of the geometry increases, its lift and drag characteristics become undesirable.

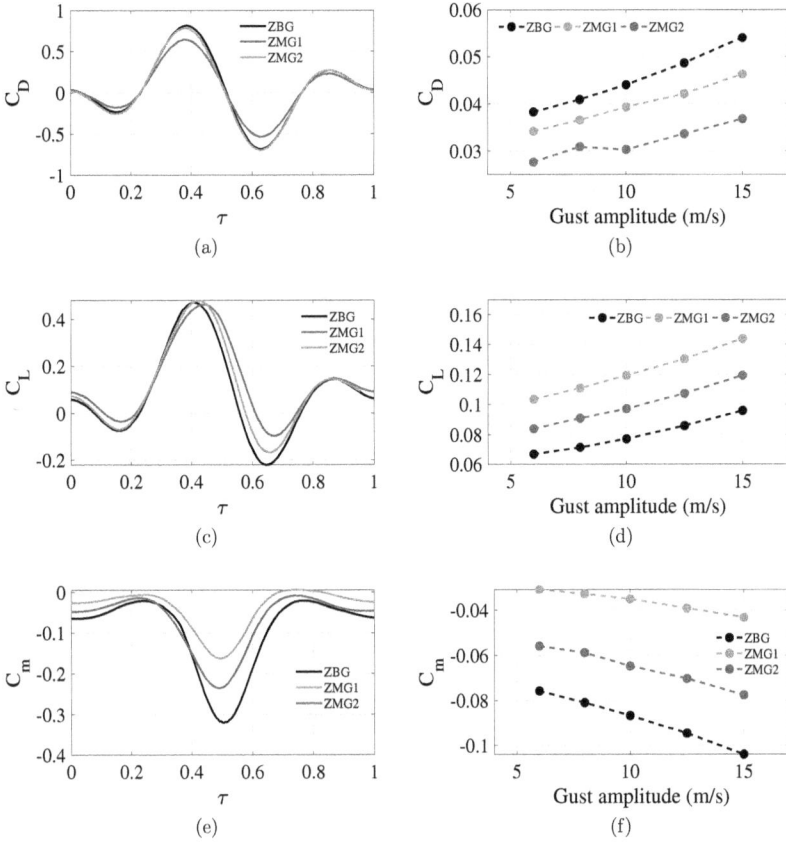

Figure 4.14: Aerodynamic analysis: Effect of aerostat shape; plots for 12.5 m/s gust - (a) drag, (c)lift and (e) moment; mean values - (b)drag, (d) lift and (f) moment.

4.2.5 Effect of gust shape

Figure 4.15 (a) shows the response for Gust 1 specified in Equation (3.1). Three different gusts are considered in this case, with different amplitudes of 6 m/s, 12.5 m/s and 15 m/s. Gust 1 possesses a symmetric profile whose rising time and falling time are the same. Symmetry in the response of drag, lift and moment is evident, even though slight variations in the peak values exist. Figure 4.15 (b) shows the Gust 2 response. Gust 2 is also a symmetric gust which has a sinusoidal nature. The response also has symmetry with variations in the peak values.

Figure 4.15 (c) represents the response for Gust 3, which is a cosine gust. It is seen from Figure 3.4 (c) that the maximum amplitude among the gusts is for Gust 3. But the drag and

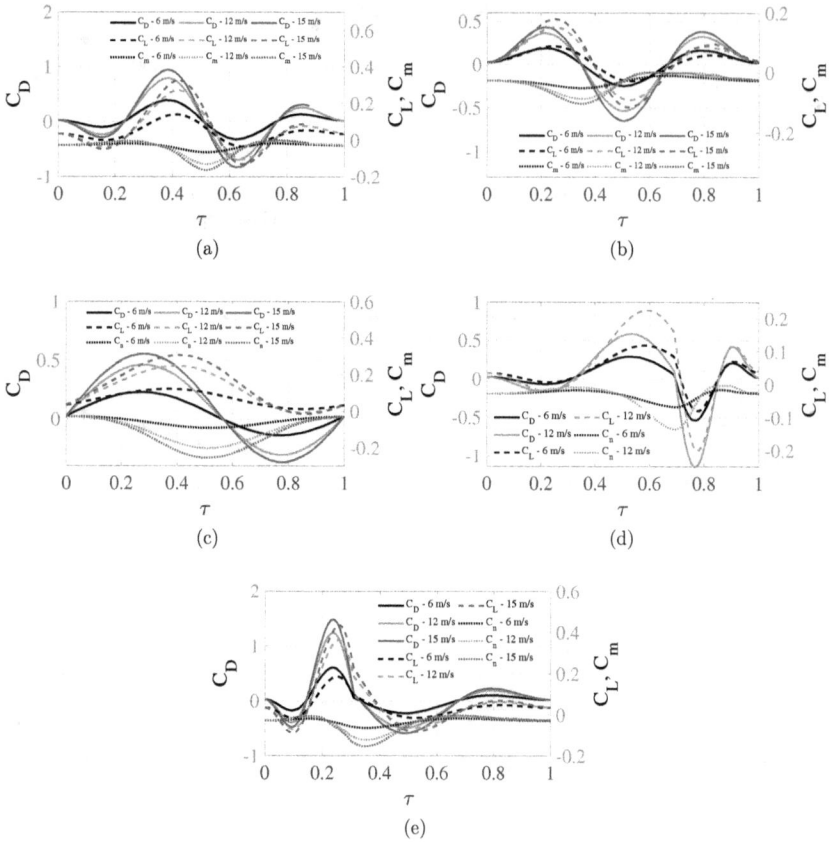

Figure 4.15: Aerodynamic analysis: Effect of gust shape; Force and moment responses for (a) Gust 1, (b) Gust 2, (c) Gust 3, (d) Gust 4 and (e) Gust 5.

lift values are not much higher for Gust 3. This is due to the smoothness of the gust, which gives it the largest rising time and falling time among the five gusts. The drag for the 15 m/s IEC gust is very much higher than that for the 15 m/s 1-cosine gust because of the sudden rise in the IEC gust. Figures 4.15 (d) and (e) show the response of the aerostat for Gust 4 specified in Equation (3.6) and Gust 5 specified in Equation (3.7), respectively. It is a modified IEC gust with different rising and falling times. Thus, the symmetric nature of the gust is lost for these two gusts. For Gust 4, the amplitude rising is very slow with a long rising time, and the falling is very fast with a short falling time and vice versa for Gust 5. Among the five gusts considered in this section, Gusts 4 and 5 have the sharpest velocity profiles, which is reflected in the response by producing the largest reported drag and lift in this research. A

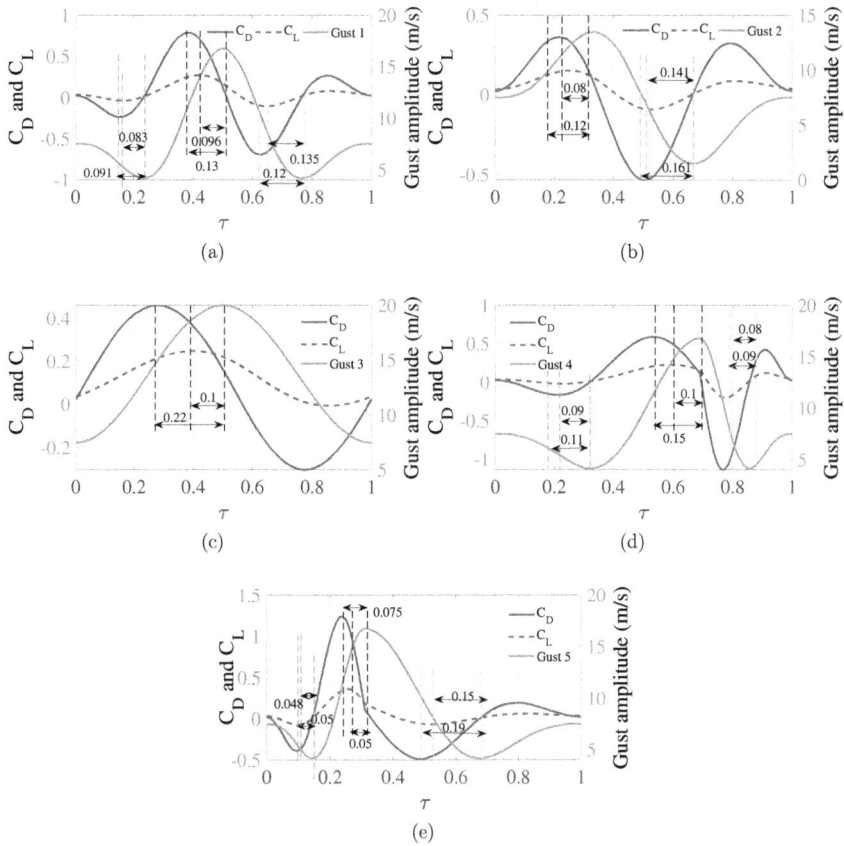

Figure 4.16: Aerodynamic analysis: Effect of gust shape - Time shifts; Comparing the temporal shift of the peak values of force responses for (a) Gust 1, (b) Gust 2, (c) Gust 3, (d) Gust 4 and (e) Gust 5.

sudden increase in the gust amplitude is predominantly affecting the aerostat response more than that of the sudden falling gust case.

Figure 4.16 shows the temporal shifts of the peaks of the force response plots relative to the corresponding peaks in the gust velocity profiles. A 12 m/s amplitude and 10.5 seconds time period gust is used for this comparison. The evolution of the amplitude of the peaks (negative and positive) of the drag, lift and moment response curves with respect to the peaks of the velocity profiles of different gusts is shown in Figure 4.17.

It can be concluded from the analysis of the effect of gust shape on the aerodynamic characteristics of the aerostat that the shape of the gust is highly affecting the gust response.

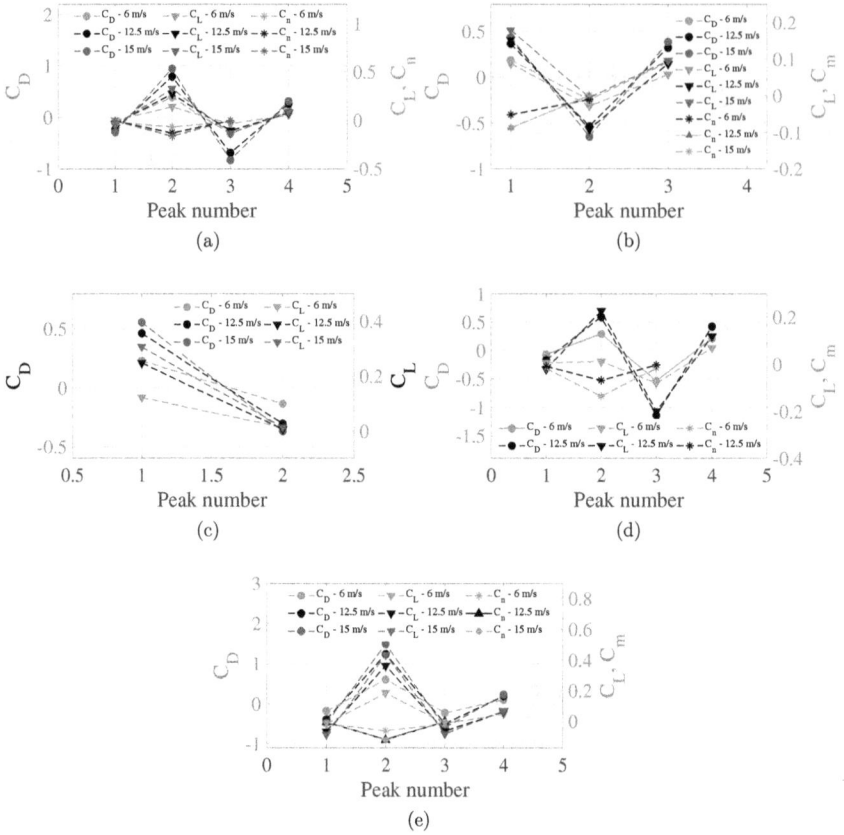

Figure 4.17: Peak amplitudes of the response for different gust shapes: Comparing the evolution of the peak values of the force responses for (a) Gust 1, (b) Gust 2, (c) Gust 3, (d) Gust 4 and (e) Gust 5.

The gusts which are having steep rising shapes are more dangerous than those with faster falling time. Even though the amplitude of the gust is higher, its effect can be less if the amplitude buildup of the gust is slow, i.e., if the gust is a smooth one.

4.2.6 Scaling of aerostat gust response

The analysis of the forces and moment about the Y-axis responses of the aerostat for different wind gust conditions pointed out a common trend in the response curves. For example, for gusts with the same gust profile and time period with different gust amplitudes, the forces and moment show similar responses with variations in the magnitudes, as shown in Figure 4.7.

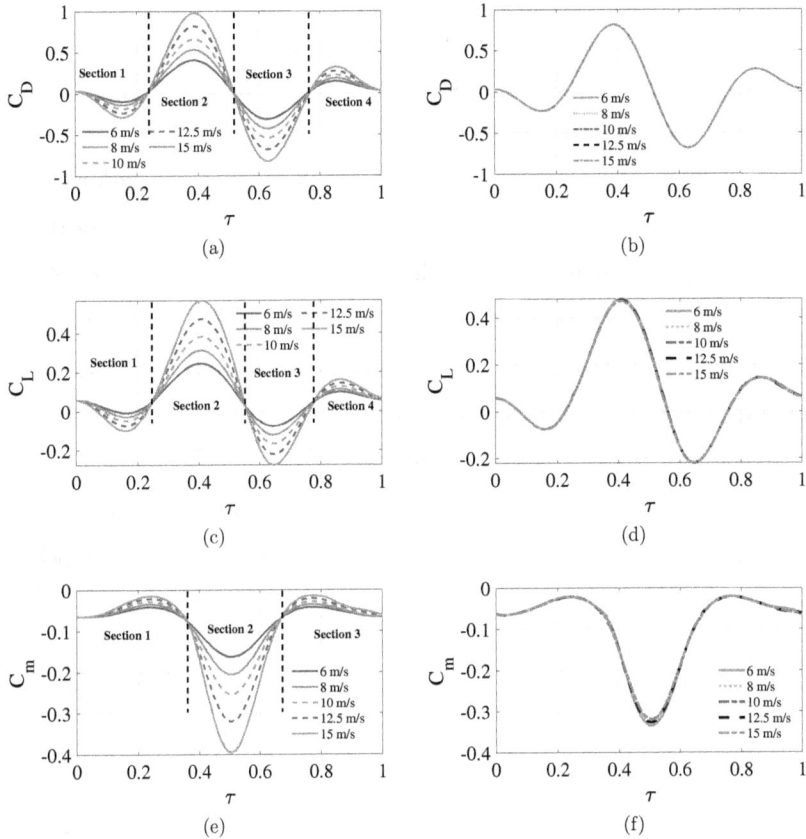

Figure 4.18: Scaling of response curves for different gust amplitudes: actual (a) drag, (c) lift, and (e) moment and scaled (b) drag, (d) lift, and (f) moment.

Table 4.3: Scaling factors used for the scaling of different gust amplitude responses.

	Section	ΔC_1	ΔC_2	ΔC_3	ΔC_4	ΔC_5	ΔC_6
C_D	1	-92.6	43.9	-0.99	-1	7E-04	-1E-05
	2	52.2	-72.6	31.4	-3.7	-0.2	0.04
	3	69.3	-259	374.8	-262.3	88.8	-11.6
	4	-78.8	356.8	-639.7	566.9	-248.2	42.9
C_L	1	-46.02	22.6	-1.03	-0.4	1.2E-03	-9E-06
	2	0.49	12.8	-22.8	13.7	-3.3	0.28
	3	110	-384.8	532.6	-363.7	122.4	-16.2
	4	96.1	-430.3	770	-688.8	308.1	-55.1
C_m	1	1.9	-1.09	-0.08	0.1	-0.005	4.6
	2	40.5	-124.6	147.2	-83.4	22.7	-2.4
	3	-16.5	65.9	-104.7	82.3	-32.03	4.9

71

The advantage of the response scaling is to represent the dynamic behaviour of the aerostat for different gust conditions using a single curve. Another advantage is that, with a proper scaling methodology, it is possible to predict the response of the aerostat for an unattended gust, such as a gust with different amplitude or AoA.

A scaling methodology is proposed here for different gust amplitudes, AoA and gust duration. A 5th degree polynomial representation for the response curves is used for the scaling analysis. Each response curve is divided into different sections for the convenience of scaling.

First, the gust cases with the same velocity profile and gust duration but different gust amplitudes are considered. The 12.5 m/s case is considered the base case here. For representing the curves using polynomials, different sections of the curves are considered. In this case, the sections are considered at the converging points in the response for drag, lift and moment, as shown in the first column of Figure 4.18. There are four sections for the drag and lift curve and three sections for the moment curve. Each section is represented using a 5th degree polynomial of the form,

$$C_1 \tau^5 + C_2 \tau^4 + C_3 \tau^3 + C_4 \tau^2 + C_5 \tau + C_6 = 0$$

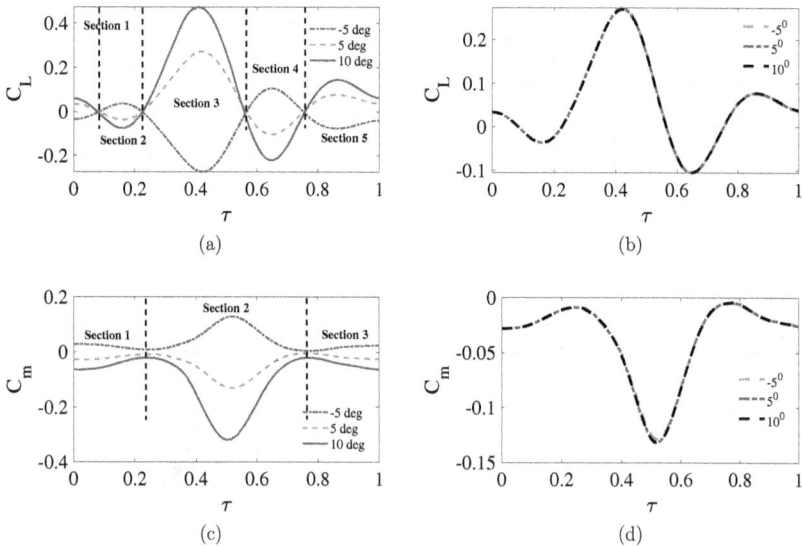

Figure 4.19: Scaling of response curves for different AoA: actual (a) lift and (c) moment, and scaled (b) lift and (d) moment.

Table 4.4: Scaling factors used for the scaling of different AoA responses.

	Section	ΔC_1	ΔC_2	ΔC_3	ΔC_4	ΔC_5	ΔC_6
	1	442.3	-57.3	5.3	-1.2	3E-04	0.005
	2	-63.1	21.7	6.1	-2.2	0.07	0.003
C_L	3	-40.8	107.6	-105.9	47.6	-9.6	0.7
	4	80.9	-310	460	-330.6	115.2	-15.5
	5	11	-30.3	24.5	-0.2	-7	2.06
	1	28.9	-21.4	4.5	-0.18	0.002	-0.007
	2	-375	775	-623.2	243.4	-46.3	3.4
C_m	3	0.6	19.3	-54.6	54.6	-23.5	3.7
	4	60.6	-265	462.6	-403.3	175.6	-30.5

The scaling can be done by representing each coefficient, C_i, using the following formulation,

$$C_i \Rightarrow C_i + 2\Delta C_i (A_b - A_k) \tag{4.1}$$

where $i = 1, ..., 5$, $k = 1, ..., 5$, C_i is the coefficient of the corresponding polynomial, A_b is the amplitude of the base case (12.5 here), and A_k is the amplitude of each gust case. ΔC_i is the scaling factor which can be calculated as,

$$\Delta C_i = \frac{(C_{i12.5} - C_{ik})}{(A_b - A_k)}$$

The scaled gust responses for different gusts with five different amplitudes are shown in the second column of Figure 4.18. Scaling factors used for the scaling of aerostat gust response for different gust amplitudes are given in Table 4.3.

In a similar manner, the scaling of the gust responses for different AoA and gust duration can be obtained. For the case of a different AoA, the 5 degree case is taken as the base case. The response curves are considered as different sections, as shown in the first column of Figure 4.19. Since the response for drag is invariant with respect to the AoA, lift and moment responses were considered for the scaling. For the AoA case, ΔC_i can be calculated as,

$$\Delta C_i = \frac{(C_{i5} - C_{ik})}{(A_b - A_k)}$$

where A_b is the base case AoA (5 degree) and A_k is the AoA of other cases.

The scaled response are shown in the second column of Figure 4.19. Scaling factors used

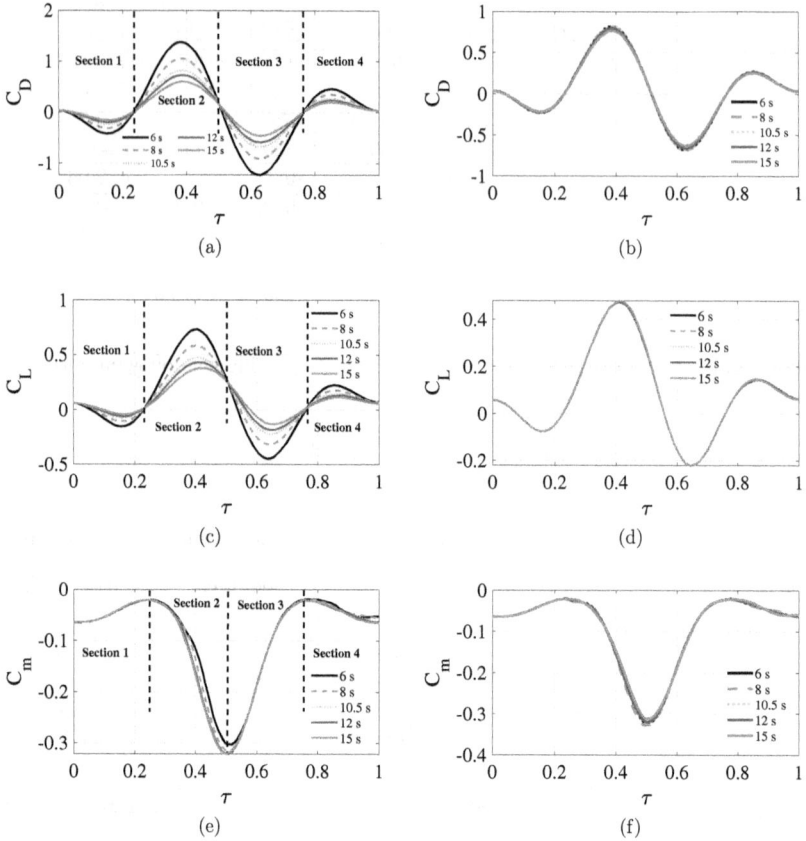

Figure 4.20: Scaling of response curves for different gust duration: actual (a) drag, (c) lift, and (e) moment and scaled (b) drag, (d) lift, and (f) moment.

for the scaling of aerostat gust response for different AoA are given in Table 4.4.

The scaling of the responses for different gust duration is done using the 10.5 seconds case as the base case. Initially, the response curves are normalised to a common X-axis. Different sections considered for the scaling are shown in the first column of Figure 4.20. Since the time duration case polynomials show a non-linear variation in their coefficient values, a slightly modified form of Equation (4.1) is used. A correction factor is included in Equation (4.1), which is the ratio of ΔC_i to the base case ΔC_i. The modified expression is as follows,

$$C_i \Rightarrow C_i + 2\frac{\Delta C_i}{\Delta C_{ib}}(A_b - A_k)$$

Table 4.5: Scaling factors used for the scaling of different gust duration responses.

	Section	ΔC_1	ΔC_2	ΔC_3	ΔC_4	ΔC_5	ΔC_6
C_D	1	132.2	-61.4	0.36	1.6	-0.001	-3E-06
	2	-69.5	83.8	-23.4	-4.2	2.2	-0.2
	3	-71.6	273.1	-398.4	278.3	-93	11.9
	4	144.7	-655.9	1177.5	-1045.9	459.4	-79.7
C_L	1	76.9	-36.5	0.7	0.8	2E-04	-1E-05
	2	-236.9	419.6	-287.1	95.8	-15.9	1.06
	3	-61.8	221.2	-308.2	208.2	-68	8.5
	4	195.9	-874	1550.7	-1367.3	598.9	-104.2
C_m	1	3.1	-2.02	0.4	-0.02	1.64	3E-07
	2	-244.5	463.6	-343.7	124.5	-22.02	1.5
	3	-14.2	45.3	-57.1	35.3	-10.7	1.2
	4	152.01	-672.5	1186.5	-1043.7	457.7	-80.07

where A_b is the base case gust duration and A_k is the gust duration for other cases.

The scaled responses are shown in the second column of Figure 4.20. Scaling factors used for the scaling of aerostat gust response for different gust duration are given in Table 4.5.

The proposed scaling methodology has been extended for the responses obtained for the IEC gust for three different aerostat geometries. An IEC gust with an amplitude of 8 m/s and a duration of 10.5 seconds was used for the analysis. The Zhiyuan geometry response was taken as the base case for the scaling. The scaling was done by considering five different sections for the drag and lift response curves and four sections for the moment curve.

For scaling the drag responses, the scaling formulation given in Equation (4.1) was modified

Table 4.6: Scaled coefficients used for ZMG1 aerostat shape responses.

	Section	ΔC_1	ΔC_2	ΔC_3	ΔC_4	ΔC_5	ΔC_6
C_D	1	-1453	704.5	-22.8	-14.8	-3E-03	0.03
	2	1313	-1931	989.2	-211	19.1	-0.7
	3	-1636.7	4176.2	-4084.5	1903.7	-423	36.4
	4	1303.6	-4607.3	6360.4	-4269.6	1387.8	-174
	5	-1764.6	8069.4	-14651.2	13197.5	-5896.1	1045
C_L	1	-399	240.2	-11.5	-6.09	-1E-03	0.06
	2	104.7	-199.06	103.7	-12.4	-1.36	0.2
	3	-1273	3487.4	-3722.9	1931.5	-487.4	48.2
	4	347.3	-1122.5	1335.9	-682.9	116.7	5.75
	5	-3300	15449	-28866	26907	-12513	-2322
C_m	1	290.2	-210.8	46	-2.8	0.12	-0.07
	2	127.96	-196.23	131.05	-50	10	-0.8
	3	89.6	-190	98.2	39.1	-45.9	9.75
	4	117.8	-599.5	1197.1	-1175.7	568.8	-108.5

as follows,

$$C_i \Rightarrow C_i \times 0.25 \left(\frac{f_b}{f_i} + (f_i - f_b) \right)$$

where f_b is the fineness ratio of ZBG and f_i is the fineness ratio of ZMGi ($i = 1, 2$). The scaling of the lift and moment responses involves both time and amplitude scaling. The time shift in the response with respect to the base case was used for the time scaling. The amplitude scaling was done using the following formulations for lift and moment responses, respectively,

$$C_{i,lift} \Rightarrow (C_i - a) \times b \tag{4.2}$$

$$C_{i,moment} \Rightarrow C_i \times c \tag{4.3}$$

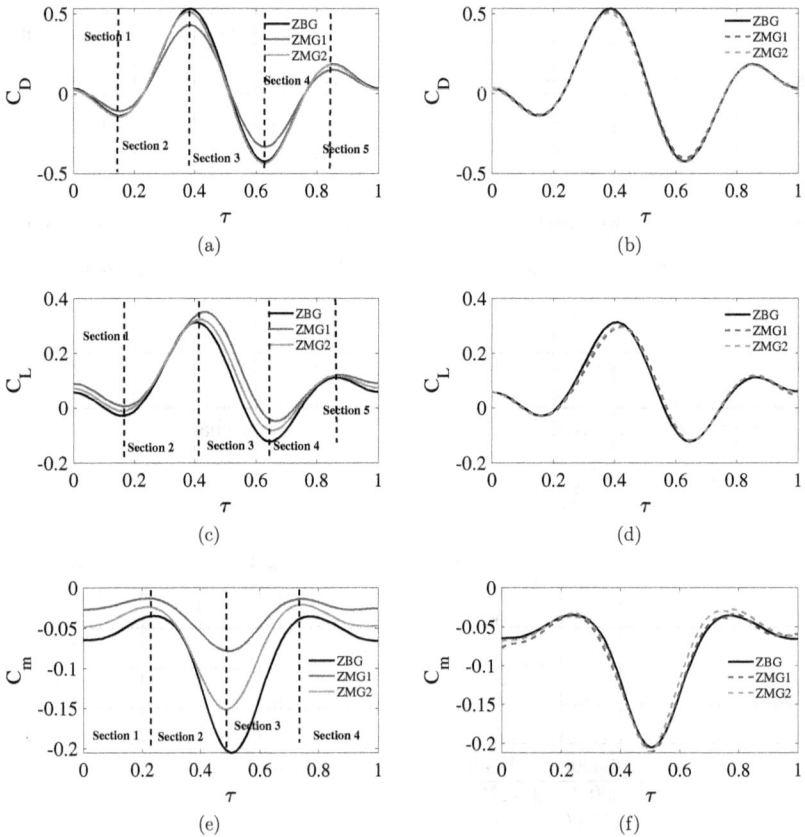

Figure 4.21: Scaling of response curves for different aerostat shapes: actual (a) drag, (c) lift, and (e) moment and scaled (b) drag, (d) lift, and (f) moment.

where a and b are 0.03 and 1 for sections 1 to 3 and 0.01 and 1.5 for sections 4 to 5 and c is 2.7.

The actual responses are shown in the first column, and the scaled responses are shown in the second column of Figure 4.21. Scaled coefficients used for the scaling of aerostat gust responses for different aerostat shapes are given in Table 4.6.

4.3 Summary

In this chapter, the effect of unsteady wind gusts on the aerodynamic characteristics of an aerostat was investigated. The unsteady conditions in the intended aerostat operating region were included as wind gusts in the simulation. Wind data from the Gujarat coastal area was used for the gust modelling.

The effect of the AoA, gust amplitude, gust duration and gust length on the aerodynamic performance of the aerostat was investigated using different gust conditions. It was found that the variation in the AoA was not affecting the aerostat drag very much. The minimum drag obtained was for the zero degree AoA.

The effect of gust length was investigated by using a wide range of time periods. It was found that as the gust length increased, the impact it made on the aerostat performance got reduced. The overshoots in the force responses diminish as the gust length increases. A comparatively smooth response was obtained for drag and lift forces. On the other hand, a slight increase in the overshoots was observed for the moment response.

From the simulation studies conducted on the aerostat model with different wind gusts, a common trend was observed among the responses. Thus, an investigation on the scaling of the aerostat responses for different AoA, gust amplitudes, aerostat shapes and gust duration was conducted. A scaling methodology was presented in which the gust responses were scaled to represent them as a single curve.

Chapter 5

Stability Derivatives of the Aerostat

5.1 Introduction

Being an aerial vehicle, the aerodynamic part of the model has excellent importance in aerostat dynamics. From the literature, it is seen that the stability derivatives based aerodynamic model is considered accurate. Stability derivatives are a measure of the degree of variation of a function or a parameter with respect to input or another parameter. Stability derivatives can be classified as velocity derivatives and acceleration derivatives (added mass terms). The derivatives due to acceleration are usually neglected for the dynamics of the heavier-than-air vehicles because of their significantly small values. The stability derivatives due to the velocity of the motion variables are usually considered along with the aerodynamic force terms in the dynamic model [25]. These terms are generally called stability derivatives, and those due to the acceleration terms are called the added mass terms.

The estimation or extraction of the stability derivatives was reported mainly using flight tests, wind tunnel tests and numerical methods. Even though the physical realism of the experimental techniques such as flight tests and wind tunnel tests is highly appreciable, they have some limitations, such as scaling errors, blocking effects and high cost. Most of the limitations of the experimental method, such as the wind tunnel tests, can be overcome by the CFD-based methods. This chapter devotes itself to presenting the extraction and analysis of the stability derivatives of the aerostat under steady and unsteady conditions. The steady condition is done by considering a constant velocity wind, and the unsteady condition is considered by a varying wind along with the aerostat in motion.

Table 5.1: Steady condition - Consolidated simulation details.

Simulation	Shape	Gust	AoA (deg)	Amplitude	Time period (s)
Grid independence	Zhi	IEC	10	12.5 m/s	10.5
Time step independence	Zhi	IEC	10	12.5 m/s	10.5
Grid motion validation	NACA-0012	Sine	0.5 to 5	2.4 deg	0.12
Method validation	Spheroid	Sine	0	1 m/s	1.6
Acceleration independence	Zhi	Sine	10	0.5, 0.75, 1 m/s	3
Frequency independence	Zhi	Sine	10	5 deg	2.7, 3, 3.3
Surge motion	Zhi, HAA, GNVR, NPL	Sine	-5, 0, 5, 10	1 m/s	3
Heave motion	Zhi, HAA, GNVR, NPL	Sine	-5, 0, 5, 10	1 m/s	3
Pitch motion	Zhi, HAA, GNVR, NPL	Sine	-5, 0, 5, 10	5 deg	2.7, 3, 3.3

5.2 Stability derivatives: Steady condition

The simulation details for each case considered for the steady condition are consolidated in Table 5.1 for better understanding.

5.2.1 Static aerodynamic characteristics

Before proceeding with the dynamic stability derivatives extraction, the static aerodynamic characteristics of the Zhiyuan aerostat, along with the other three aerostats, were studied. It was done by conducting transient simulations with a steady velocity of 7.5 m/s for each aerostat. There are no oscillations induced on the aerostat for the static derivative extraction simulations.

A comparison of the drag and lift for the Zhiyuan base case and the other three aerostats for different AoA are shown in Figure 5.1 (a) and (b). The symmetric nature of the variation of the forces with the AoA is clearly seen in the figure. The axisymmetric nature of the aerostat shapes causes these symmetric characteristics. The highest drag is shown by the Zhiyuan shape for all the AoA. It is due to the larger front area and the larger surface area of the shape among all four aerostat shapes. They contribute to the pressure as well as the skin friction drag. As the AoA increases from zero, the area exposed to the free stream wind increases and aids the drag formation. The drag response for the other three aerostats is also shown in Figure 5.1 (a). All the other three aerostats also have similar drag characteristics with a reduced magnitude. NPL shape has the least drag among the four aerostat shapes because of its least frontal area and surface area. GNVR aerostat and HAA aerostat have the same drag characteristics. This is due to the similar leading edge geometry for those two aerostats. The Zhiyuan aerostat has a comparatively non-aerodynamic leading edge, as seen in Figure 3.3, which causes the higher drag response among the four aerostats.

As the aerodynamic lift is generated by the differential pressure distribution around the aerostat surfaces, it also shows variation with the AoA. The lift response of the Zhiyuan aerostat, along with the other three aerostats, is shown in Figure 5.1 (b). Zhiyuan shape has the maximum lift among all the four shapes. As the AoA increases, the lift magnitude increases largely. The large volume of the Zhiyuan aerostat has a higher differential pressure between the top and bottom surfaces, thereby, the higher lift. As seen for the drag response, the lift response is also similar for the other three aerostats. NPL aerostat has the least lift magnitude, and GNVR aerostat has the maximum lift among the three aerostats.

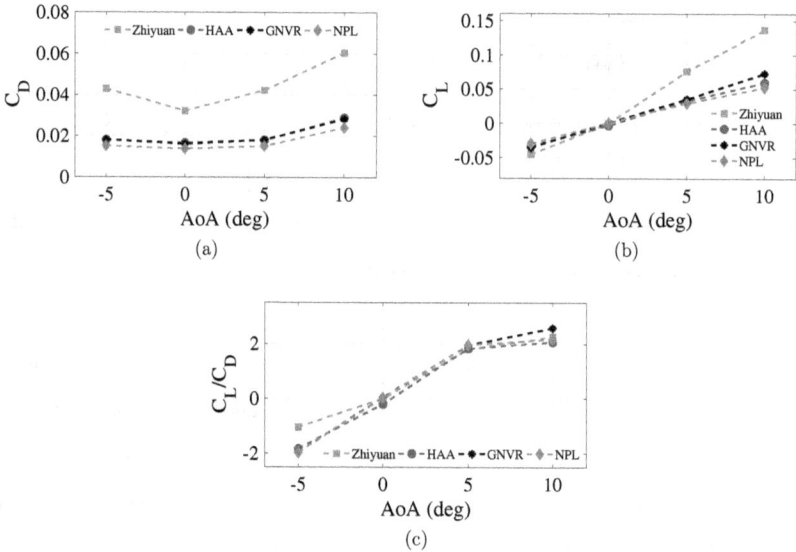

Figure 5.1: Static aerodynamic coefficients of four aerostats for different AoA: (a) coefficient of drag, C_D, (b) coefficient of lift, C_L and (c) lift-drag ratio.

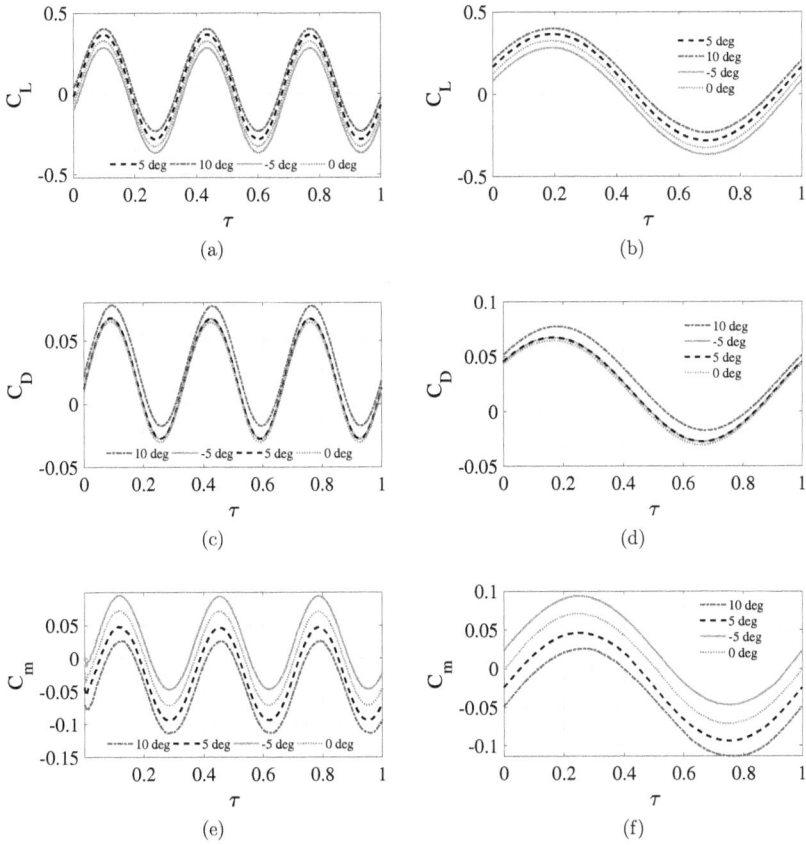

Figure 5.2: Zhiyuan aerostat response for (a) heave oscillation - 3 cycles, (b) heave oscillation - 1 cycle, (c) surge oscillation - 3 cycles, (d) surge oscillation - 1 cycle, (e) pitch oscillation - 3 cycles and (f) pitch oscillation - 1 cycle.

The primary criteria for the selection of the shape for an aerostat are the reduced drag and maximum lift for a particular flight condition. A low drag profile may result in a lower lift and vice versa. Therefore, the aerodynamic shape for the aerostats is selected based on the lift-drag ratio and not on the individual lift and drag information. Figure 5.1 (c) shows the lift-drag ratio of the four aerostats for different AoA. It is seen that all four aerostats exhibit similar characteristics except for Zhiyuan at the negative AoA. This justifies the authors' selection of these aerostat shapes for the analysis. The ultimate goal is to select a low drag shape with maximum lift for the wind measurement application.

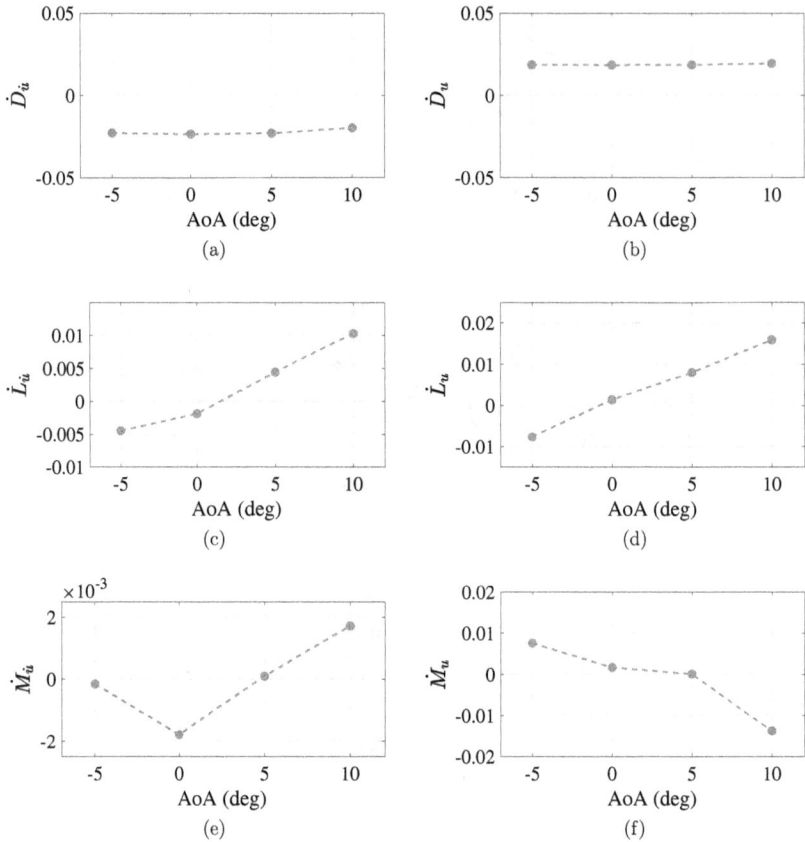

Figure 5.3: Zhiyuan stability derivatives for surge oscillations at different AoA.

5.2.2 Dynamic stability derivatives

The dynamic stability derivative extraction methodology presented in Chapter 3 was applied to the four aerostats with four different AoA. Three full cycles of responses were obtained for all the simulation cases, and a full stable cycle was selected for the analysis. The heave and surge oscillations were simulated using an amplitude of oscillation of 1 m/s and a frequency of oscillation of 2.09 rad. The pitch oscillations were simulated using an amplitude of oscillation of 5 deg and a frequency of oscillation of 2.09 rad, 2.32 rad and 1.9 rad. For each aerostat shape, the drag, lift and moment responses were measured for the heave, surge and pitch oscillations. All the responses are presented with non-dimensional time on the X-axis.

Table 5.2: Validation of the Zhiyuan stability derivative.

Stability derivative	Current study	Wang [78]	% deviation
$\dot{M}_{\dot{q}}$	0.0032	0.003399	4.98

Zhiyuan aerostat

The response of the Zhiyuan aerostat for heave oscillation, surge oscillation and pitch oscillation are shown in Figure 5.2 (a) and (b), Figure 5.2 (c) and (d), and Figure 5.2 (e) and (f), respectively. Lift response for the heave oscillation, drag response for the surge oscillation and moment response for the pitch oscillation with 3 s time period are shown in the figure.

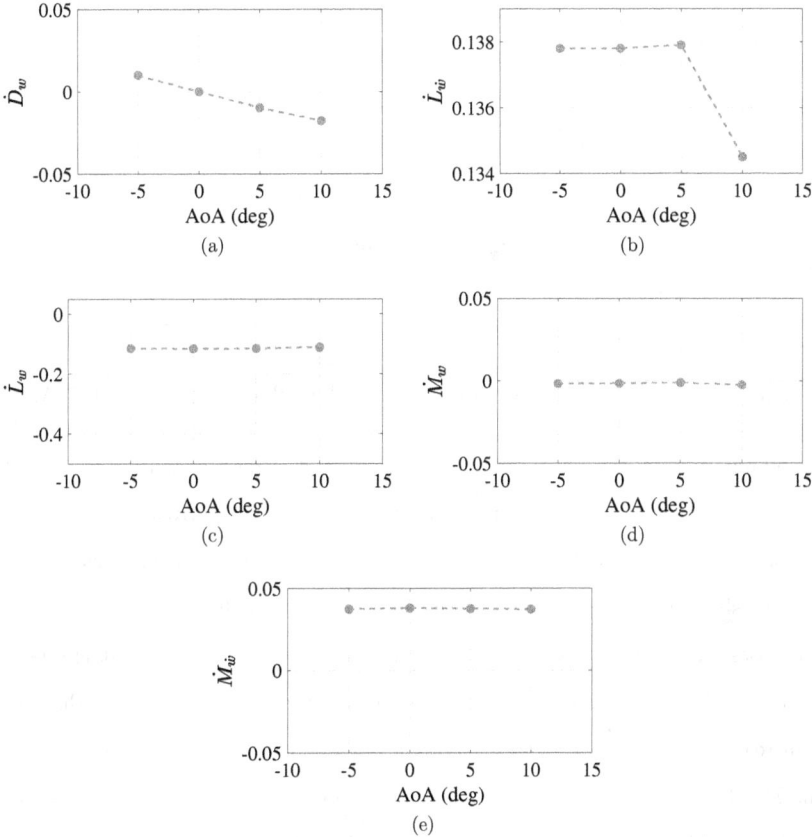

Figure 5.4: Zhiyuan stability derivatives for heave oscillations at different AoA.

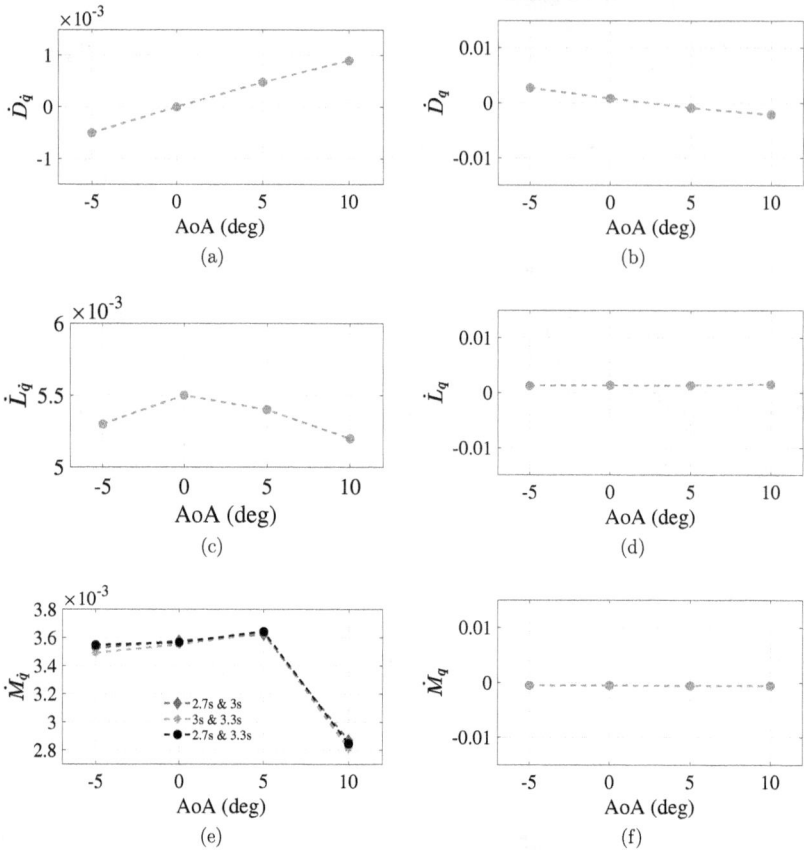

Figure 5.5: Zhiyuan stability derivatives for pitch oscillations at different AoA.

The first column of the figure represents a continuous set of responses, and the second column represents the single-cycle plots that are used for the stability derivatives extraction. As the input oscillations are sinusoidal, the responses are also sinusoidal, as shown in the figure.

As previously seen from the static aerodynamic response of the aerostat, the magnitude of lift increases with the increase in the AoA, as shown in Figure 5.2 (a). It is due to the variation in the differential pressure between the top and the bottom surfaces of the aerostat. Further increase in the AoA may cause stall conditions, and it is not in the scope of the current research. The drag remains constant for the smaller AoA, as shown in Figure 5.2 (c), because the frontal area hit by the wind remains almost unchanged. As the AoA increases to 10 degrees, the frontal area of the aerostat increases and the drag increases. This is also seen in the static stability derivative shown in Figure 5.1 (a). The moment response shown in Figure 5.2 shows an inverse relation with the AoA. The moment reduces as the AoA increases and vice versa.

The stability derivatives of the Zhiyuan aerostat required for the longitudinal dynamic model were extracted using the methodology specified in Chapter 3. The stability derivatives obtained for the Zhiyuan aerostat were compared and validated with the results available in the open literature, as shown in Table 5.2. The moment derivative due to pitch acceleration ($M_{\dot{q}}$) result was compared, and the deviation was less than 5 %.

A comparison of the stability derivatives extracted from the surge oscillations, heave oscillations and pitch oscillations for different AoA are shown in Figures 5.3, 5.4 and 5.5, respectively. Drag derivative due to axial acceleration and velocity shown in Figure 5.3 (a) and (b) respectively show almost constant values for the small AoA. As the AoA increases to 10 degrees, the derivative increases as well. This is in close agreement with the drag response shown in Figure 5.2 (a). Lift derivative due to the axial acceleration and velocity are shown in Figure 5.3 (c) and (d), respectively, showing a linear variation with the AoA. This shows the peculiar behaviour of the aerostat to build up the lift as the axial acceleration happens, which aids the dynamic stability of the aerostat. The magnitude of the derivatives shifts from negative to positive as the AoA changes from negative to positive. At the zero AoA, the derivative value is close to zero. That means the acceleration of the aerostat at zero degree AoA will not affect much on the lift. Moment derivatives due to axial acceleration and velocity shown in Figures 5.3 (e) and (f), respectively, showing a linear decrease with the AoA. There is an exception at a 10-degree AoA for the acceleration derivative. Thus, the acceleration along the axial direction causes the moment to decrease, which will reduce the control effort for pitch control.

The drag derivative due to the vertical velocity shown in Figure 5.4 (a) is linearly decreasing with the AoA, whereas the lift derivative is almost constant, as shown in Figure 5.4 (c). The lift derivative due to vertical acceleration shown in Figure 5.4 (b) remains constant for small AoA but reduces to a lower value for the higher AoA. The vertical acceleration mainly caused by vertical gusts will not aid the drag buildup of the aerostat. On the other hand, it will not even aid the lift generation of the aerostat. The moment derivative due to vertical velocity shows a variation opposite to that of the moment derivative due to axial acceleration, as shown in Figure 5.4 (c). The moment derivative due to vertical acceleration remains constant for all the AoA, with an exception at zero degree, as shown in Figure 5.4 (d).

The drag, lift and moment derivatives due to the pitching acceleration shown in Figure 5.5 (a), (c) and (e), respectively, include the application of two different oscillations as explained in Equation (3.24) with time periods of 2.7, 3 and 3.3 seconds. Three different sets of results were presented to check the effect of the selection of a second frequency for the calculation. The results show considerable similarity among the frequency pairs. For the remaining aerostat shapes, the oscillation frequencies for the pitching oscillations were taken as 2.7 seconds and 3 seconds. The drag derivative, due to the pitching velocity, shows a linear decrease with the AoA, whereas the lift derivative shows an increase with the AoA. Moment derivative due to the pitching velocity remains almost constant with the AoA variation. The pitching acceleration aids the lift generation and weakens the drag generation of the aerostat.

GNVR, HAA and NPL aerostats

The simulations for extracting the stability derivatives for the longitudinal dynamic model of the aerostat by inducing surge, heave and pitch oscillations were repeated for the GNVR, HAA and NPL aerostats as well. Here in this section, some of those results which showed deviation from the Zhiyuan response are presented.

The response of the GNVR aerostat, HAA aerostat and NPL aerostat for heave oscillation, surge oscillation and pitch oscillation is shown in Figure 5.6. A stable cycle of response used for the stability derivatives extraction is shown in the figure.

The lift response of the GNVR aerostat for different AoA shows an increment with the AoA as seen similar to the static stability derivatives, as shown in Figure 5.6 (a). The close similarity between the Zhiyuan shape and GNVR shape made the lift response of these two aerostats similar. The drag response of the GNVR aerostat shown in Figure 5.6 (b) has a considerable deviation from that of the Zhiyuan aerostat. Due to the more streamline design, the Zhiyuan aerostat has a reduced drag compared to the GNVR aerostat. The moment response of the GNVR aerostat shown in Figure 5.6 (c) is similar to that of the Zhiyuan aerostat. The lift response of the HAA aerostat for different AoA is shown in Figure 5.6 (d). The amplitude of the lift increases with the increase in the AoA. The drag response shown in Figure 5.6 (e) demonstrates an invariant relation with the AoA. The moment response, shown in Figure 5.6

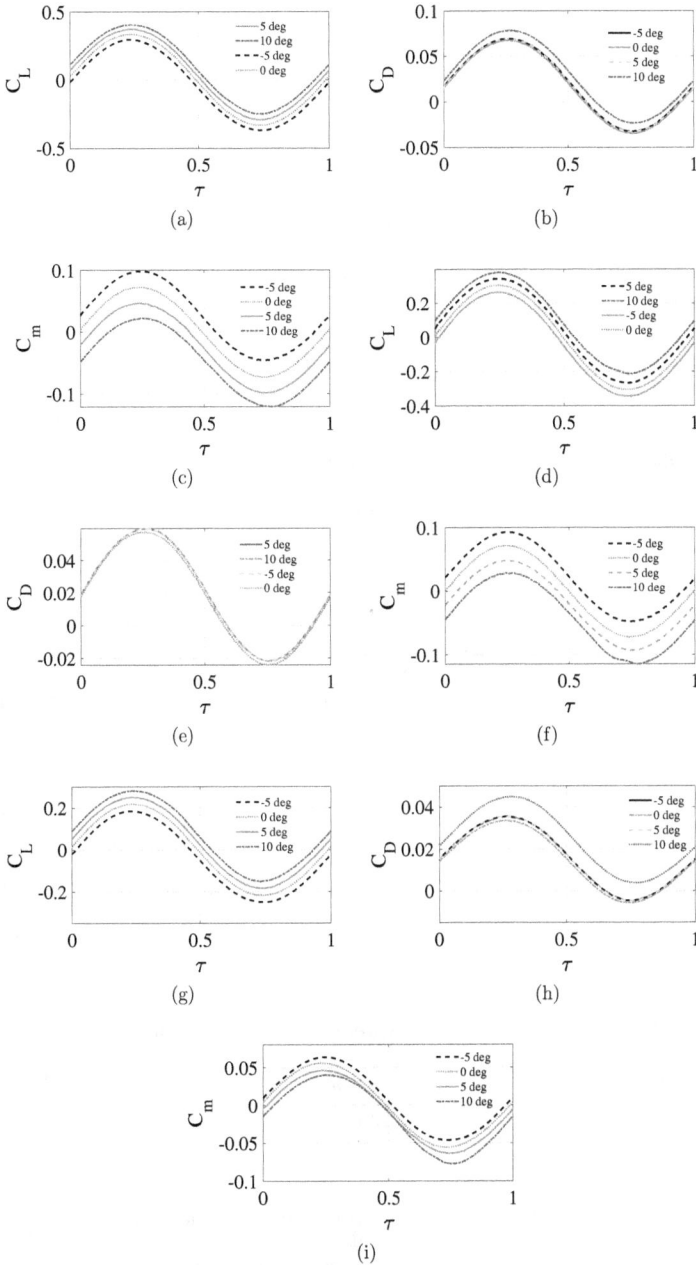

Figure 5.6: Response of GNVR aerostat for (a) heave, (b) surge and (c) pitch oscillations, HAA aerostat for (d) heave, (e) surge and (f) pitch oscillations, and NPL aerostat for (g) heave, (h) surge and (i) pitch oscillations.

(f), reduces with the increase in the AoA. The lift response of the NPL aerostat shown in Figure 5.6 (g) shows the least lift magnitude for the given AoA, which is in close agreement with the static derivative response given in Figure 5.1 (b). The drag response remains constant for the smaller AoA and shows a slight increase in the magnitude for the 10-degree AoA, as shown in Figure 5.6 (h). The drag magnitude is also the least among the four aerostats, as seen from the static derivative results given in Figure 5.1 (a). The moment response of the NPL aerostat, similar to the other three aerostats, shows an inverse relation with the AoA, as shown in Figure 5.6 (i).

The stability derivatives obtained for the Zhiyuan aerostat are compared with those for the other aerostats, as shown in Figure 5.7. The drag derivatives due to axial acceleration for the four aerostats shown in Figure 5.7 (a) remain constant with the AoA. This is in agreement with the static aerodynamic characteristics shown in Figure 5.1 (a). Even though the individual derivatives remain constant, their magnitude varies largely among the aerostats. GNVR aerostat has the maximum value, and the NPL has the minimum value, as shown in the figure. As seen for the Zhiyuan aerostat case in Figure 5.3 (a), the drag derivatives for the 10-degree AoA show a deviation for all the aerostats. It is due to the increase in the frontal area of the aerostat due to the large AoA. The magnitude of the drag derivative for the GNVR aerostat is slightly higher than that for the Zhiyuan aerostat. This can be explained by the slightly larger diameter of the GNVR aerostat, which causes the front area to increase slightly. In a similar way, the diameter of the NPL aerostat is smaller than the Zhiyuan aerostat, which causes the front area to be smaller than the other aerostat and thereby resulting in a smaller drag. This pattern of drag magnitude is reflected in the rate of change of drag as well. As the diameter of Zhiyuan and HAA aerostats were the same, their drag response and the drag derivative remained the same, as shown in the figure. The lower drag derivative value of the NPL aerostat helps to maintain the attitude of the aerostat because as the aerostat accelerates forward, the drag buildup will be comparatively less among the four aerostats. On the other hand, the other three aerostats need more control effort for station keeping.

The lift derivative due to the axial acceleration for the four aerostats is shown in Figure 5.7 (b). There is a peculiar difference between the derivative response of the Zhiyuan aerostat among the other aerostats. As the AoA increases, the rate of change of lift due to the axial

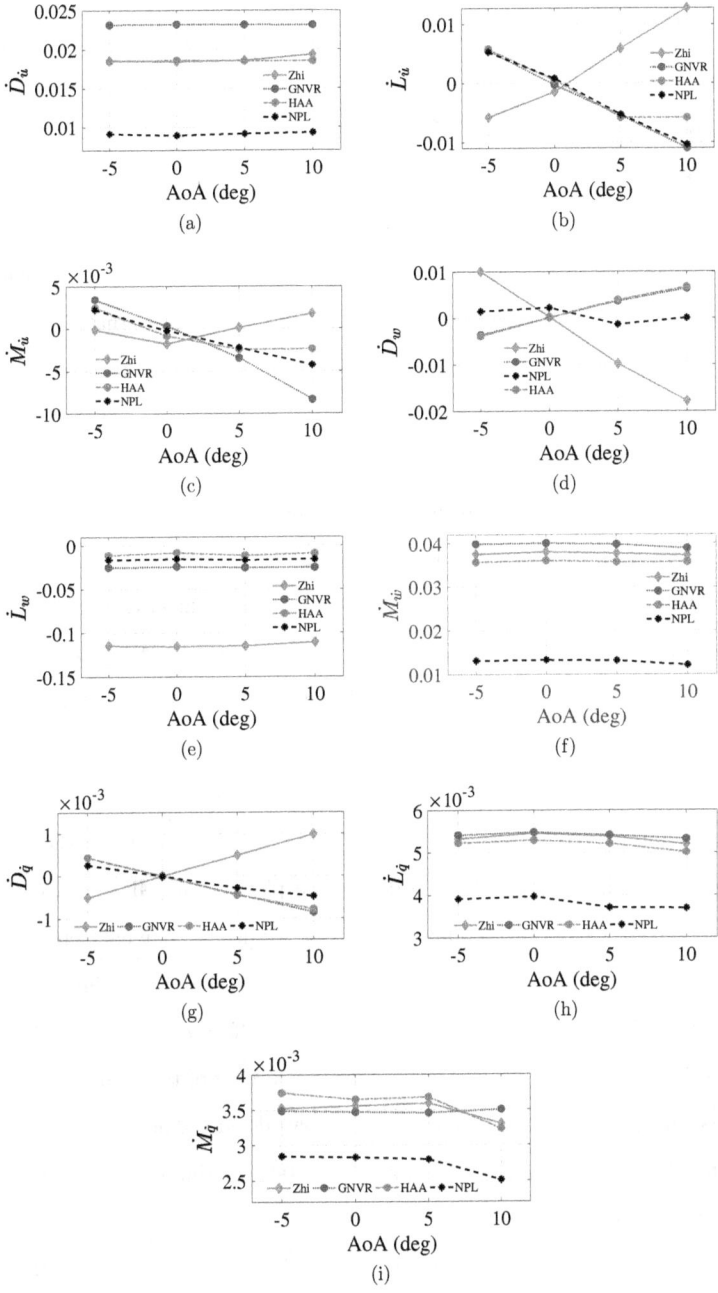

Figure 5.7: Stability derivatives comparison of four aerostats.

acceleration increases linearly, whereas the derivatives decrease for the other aerostats. This particular behaviour of the Zhiyuan aerostat is advantageous for the dynamic stability of the aerostat as the increased lift helps the aerostat to be more stable. The geometry of the Zhiyuan aerostat causes this particular response. All the other aerostats considered have a less pointy leading edge than the Zhiyuan aerostat. The positive slope in the lift derivative with the axial acceleration of the Zhiyuan aerostat aids the altitude control and station-keeping control efforts. The acceleration along the axial direction causes the lift to reduce its magnitude for the other three aerostats, unlike for the Zhiyuan aerostat case. It shows that the aerostat fails to retain its lift as its own. There should be control actions to maintain the required lift for the aerostat.

The moment derivatives due to the axial acceleration for the four aerostats are shown in Figure 5.7 (c). The rate of change of moment decreases with the AoA for the other three aerostats. The reduction in the moment derivative due to the axial acceleration will reduce the control effort for the pitch control. The Zhiyuan aerostat moment derivative increases for the positive AoA, as shown in the figure. Thus, for large positive AoA, the Zhiyuan aerostat demands more control effort for maintaining the desired pitch. Even though the stalling of the aerostat is not in the scope of this work, the pattern in the moment derivative for the Zhiyuan aerostat suggests that stalling occurs at a lower AoA than for the other three aerostats.

The drag derivatives due to the vertical velocity for the four aerostats are shown in Figure 5.7 (d). There is a positive slope for the derivatives plot for the GNVR, HAA and NPL aerostats and a negative slope for the Zhiyuan aerostat. Unlike the Zhiyuan aerostat, the drag magnitude builds up as the other three aerostats accelerate in the vertical direction, primarily due to vertical wind gusts. This demands more control effort for the station keeping and attitude control for the aerostats. This shows that the turning of the aerostat for a pitch up flight, such as taking off from the ground station, slows down the aerostat due to the extra drag build up. The negative slope of the Zhiyuan aerostat aids the control efforts to maintain the position of the aerostat.

The lift derivatives due to the vertical velocity for the four aerostats shown in Figure 5.7 (e) possess constant values for different AoA. The vertical acceleration can be due to a strong vertical wind gust or an ascending flight phase at the time of the installation or maintenance of

Table 5.3: Unsteady condition - Consolidated simulation details.

	Motion	AoA (deg)	Gust amplitude (%)	Time period (s)	Motion amplitude
Grid independence	-	10	66	10.5	-
Time step independence	-	10	66	10.5	-
Grid motion validation	Pitch	0.5 - 5	-	0.12	2.4 deg
Unsteady	Surge	-5	10, 30, 50	3	1m/s
	Surge	0	10, 30, 50	3	1 m/s
	Surge	5	10, 30, 50	3	1 m/s
	Surge	10	10, 30, 50	2.7, 3, 3.3	1 m/s
	Heave	-5	10, 30, 50	3	1 m/s
	Heave	0	10, 30, 50	3	1 m/s
	Heave	5	10, 30, 50	3	1 m/s
	Heave	10	10, 30, 50	2.7, 3	1 m/s
	Pitch	-5	10, 30, 50	2.7, 3	5 deg
	Pitch	0	10, 30, 50	2.7, 3	5 deg
	Pitch	5	10, 30, 50	2.7, 3	5 deg
	Pitch	10	10, 30, 50	2.7, 3, 3.3	5 deg

the aerostat. The Zhiyuan aerostat has the maximum lift derivative in the negative direction, which demands more control effort for station keeping. All the other aerostats aid the lift for the ascending phase, which helps to reduce the control effort. The moment derivatives due to the vertical acceleration for the four aerostats shown in Figure 5.7 (f) also remain constant for different AoA. Zhiyuan, HAA, and GNVR aerostats have the maximum value among the four aerostats, which makes the pitch control rapid as well as requires more effort to keep them in the desired pitch. The NPL aerostat has the minimum value, which helps the aerostat to vary the pitch smoothly with less control effort. The force and moment derivatives due to the pitch acceleration for the four aerostats are shown in Figure 5.7 (g) - (i). The drag derivative varies linearly with the AoA with a negative slope, whereas the Zhiyuan aerostat has a positive slope. The lift derivative shows an almost constant value for different AoA, with the NPL aerostat having the minimum value. The moment and lift derivatives show similar values as the Zhiyuan aerostat except for the NPL aerostat.

5.3 Stability derivatives: Unsteady condition

The unsteady analysis of the aerostat was done by considering various parameters such as the AoA, phase shift between the inputs, amplitude of the wind gust, and time period of the wind gust. The sinusoidal signal used for the oscillatory motion of the aerostat was kept constant for surge, heave, and pitch oscillations, respectively. The phase shift was introduced by shifting the oscillatory sine signal with respect to the wind sine signal. The simulation details for each case considered for the unsteady condition are consolidated in Table 5.3 for better understanding.

5.3.1 Effect of phase shift

The analysis of the effect of the phase difference between the wind and the aerostat oscillation was conducted by considering four phase angles: 0, 90, 180, and 270 degrees. The wind gust sinusoidal signal is taken as the reference for the phase shifting of the oscillatory sinusoidal signals.

A representative set of results for the surge oscillations with a 10-degree AoA and for a wind gust of 50 % amplitude (50 % of the mean wind velocity added to the mean wind amplitude) are shown in Figure 5.8. The drag response of the aerostat for different phase shifts is shown in Figure 5.8 (a). The effect of phase shift is reflected in the response as a time shift with respect to the 0 phase shift case. Figure 5.8 (d) shows the time shift of drag response with respect to the 0 phase shift case. A maximum shift occurs for the 180-degree case as expected because it has the maximum difference between the two sinusoidal signals, as seen in Figure 3.13. All the other three responses are leading the zero degree case. The lift response of the aerostat for different phase shifts is shown in Figure 5.8 (b). There is a time shift among the lift responses similar to the drag response, as shown in Figure 5.8 (e). The maximum shift is for the 90-degree case, and it is leading, whereas the other two responses are lagging behind the 0-degree response. The moment response of the aerostat for different phase shifts is shown in Figure 5.8 (c). Even though the response is periodic, there is more than one sinusoidal component in the response, which makes the curve have multiple crests and troughs. Among the drag, lift, and moment responses, the moment response shows the maximum shift with respect to its 0-degree case, as shown in Figure 5.8 (f). The 0-degree response leads all the other three responses, and

92

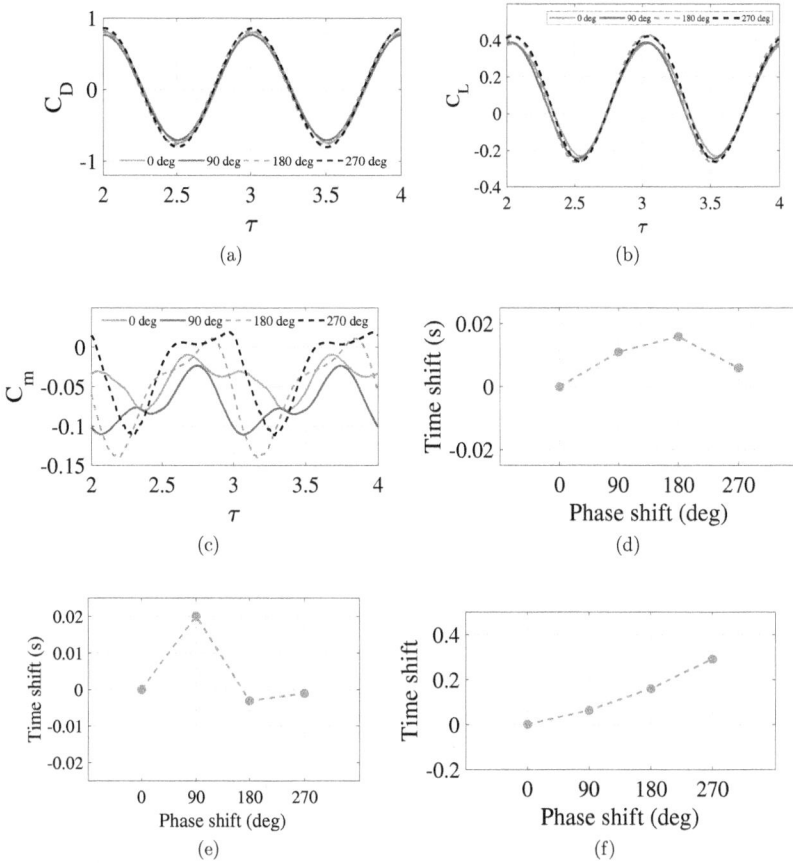

Figure 5.8: Results of the analysis of the effect of phase shift between the input signals for surge oscillation: (a) drag response, (b) lift response, (c) moment response, and time shift with respect to the 0 degree case for (d) drag, (e) lift, and (f) moment.

there is a linear increment in the time shift as the phase shift increases.

The response of the aerostat for the heave oscillations with a 10-degree AoA and for a wind gust of 10 % amplitude is shown in Figure 5.9. The drag response, shown in Figure 5.9 (a), shows a variation in amplitude and phase among the responses. The phase shift of the drag responses with respect to the 0-degree case is more compared to that of the surge case, as shown in Figure 5.9 (d). The 0-degree response lags behind all the other three responses, and the 180-degree case has the maximum phase shift from the 0-degree case. The lift response of the aerostat for different phase shifts is shown in Figure 5.9 (b). The variation in the amplitude

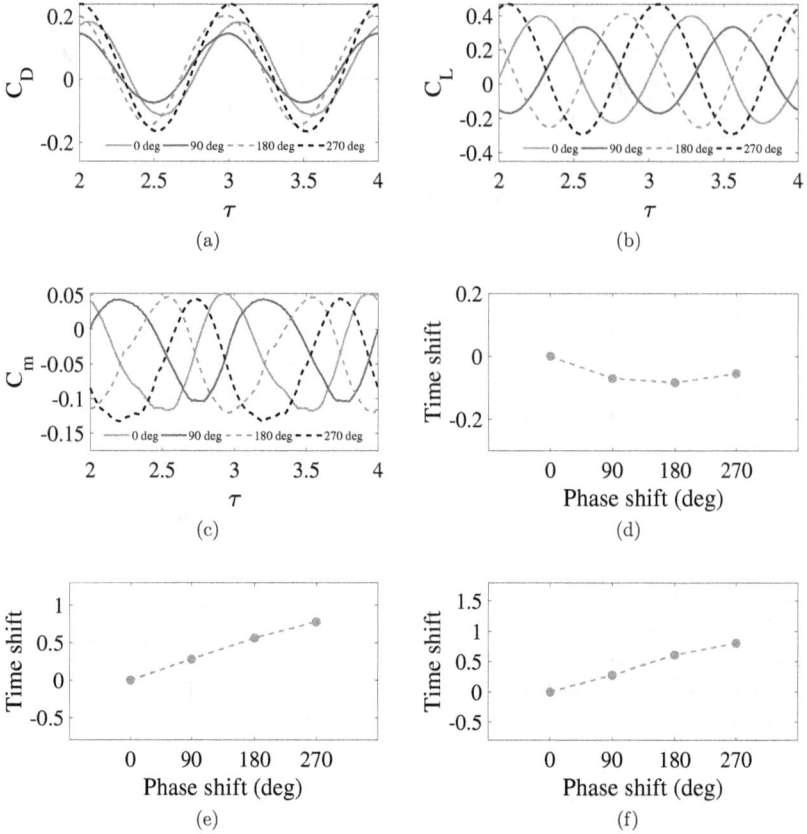

Figure 5.9: Results of the analysis of the effect of phase shift between the input signals for heave oscillation: (a) drag response, (b) lift response, (c) moment response, and time shift with respect to the 0 degree case for (d) drag, (e) lift, and (f) moment.

and phase shift is predominantly visible compared with the surge case. The phase shift among the responses is linearly increasing, as shown in Figure 5.9 (e). The 0-degree response leads all the other three responses, and there is a linear relation in the variation of phase shifts, as shown in the figure. The moment response for heave oscillations is shown in Figure 5.9 (c). Similar to the lift response, the variation in the phase is visible with a minimum variation in the amplitudes. Figure 5.9 (f) shows the phase shift among the moment responses which is linearly increasing. Similar to the lift case, the 0-degree response leads all the other three responses.

A representative set of results for the pitching oscillations of the aerostat for a 0-degree

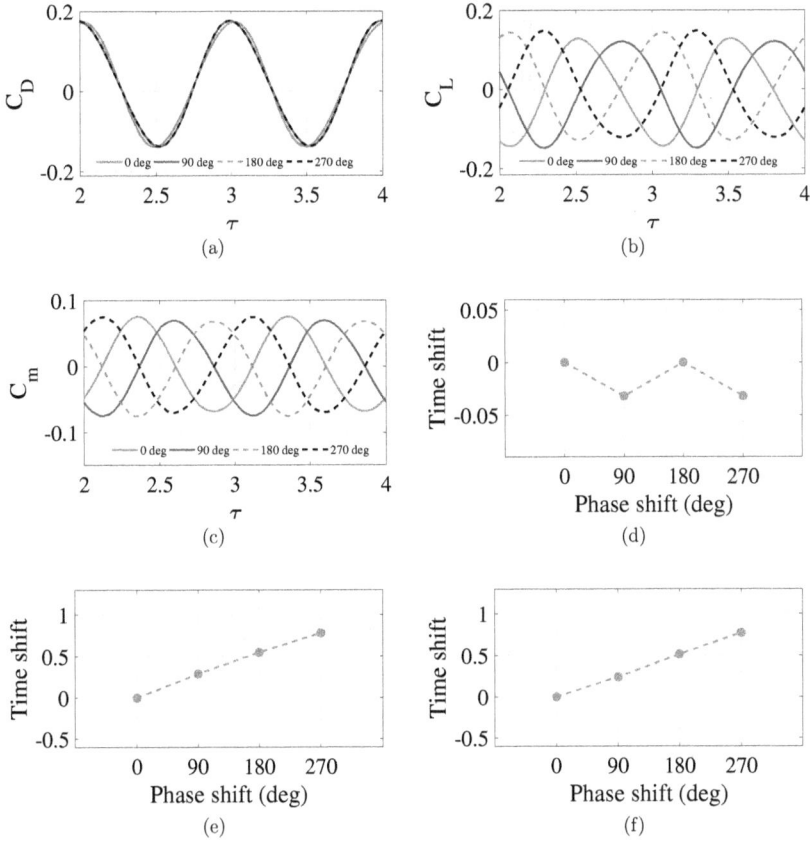

Figure 5.10: Results of the analysis of the effect of phase shift between the input signals for pitch oscillation: (a) drag response, (b) lift response, (c) moment response, and time shift with respect to the 0 degree case for (d) drag, (e) lift, and (f) moment.

AoA and 10 % gust amplitude are shown in Figure 5.10. The drag response, shown in Figure 5.10 (a), remains almost constant in terms of the amplitude and time shift. Whereas the lift and moment responses shown in Figure 5.10 (a) and (c) respectively demonstrate time shift with respect to the input phase shift. Variation in the amplitude of responses remains almost constant. The variation of drag, lift, and moment time shift with respect to the input phase shift are shown in Figure 5.10 (d), (e), and (f). The variation for the lift and moment responses are linear as well as lagging in the 0-degree case.

The variation of the amplitude of drag, lift, and moment responses of the aerostat with a phase shift between the input signals undergoing surge, heave, and pitching oscillations are

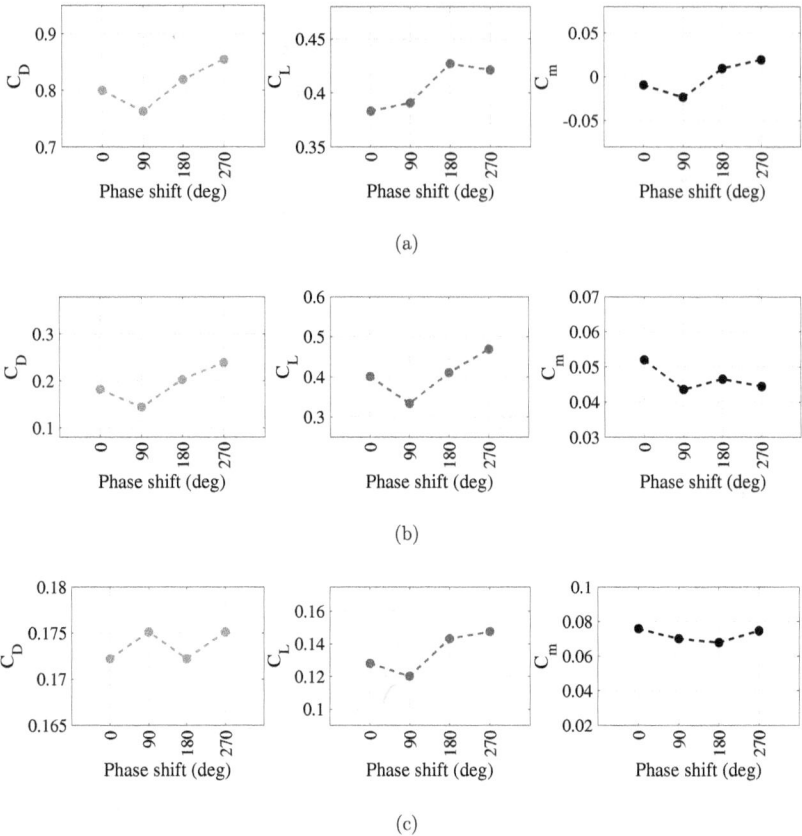

Figure 5.11: Variation of forces and moment amplitude with phase shift between inputs for (a) surge, (b) heave, and (c) pitch oscillations.

shown in Figure 5.11 (a), (b), and (c), respectively.

5.3.2 Effect of gust amplitude

For analysing the effect of wind gust amplitude, three gust conditions were considered: gust with 10 %, 30 %, and 50 % of the steady wind velocity. Figure 5.12 shows the response for surge oscillations with a 10-degree AoA and 90-degree phase shift case. The drag response for different gust amplitudes is shown in Figure 5.12 (a). The increase in the gust amplitude results in a rise in the drag response. Apart from the variation in the amplitude, there is a slight time shift in the drag response, as shown in Figure 5.12 (d). The lift response for different gust amplitudes is shown in Figure 5.12 (b). There is a considerable variation in the

96

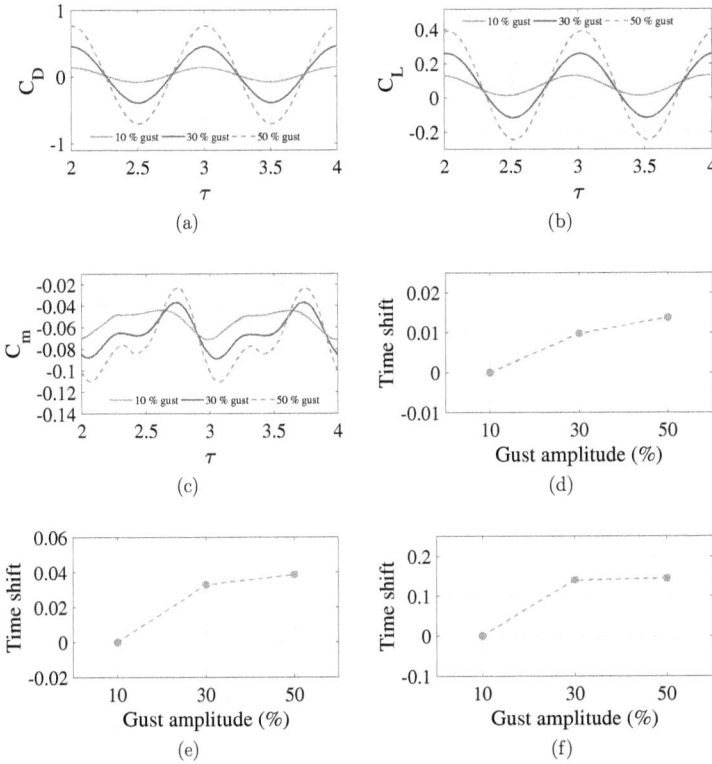

Figure 5.12: Results of the analysis of the effect of gust amplitude for surge oscillation: (a) drag response, (b) lift response, (c) moment response, time shift with respect to the 10 % case for (d) drag, (e) lift, and (f) moment.

lift response with the gust amplitude. The time shift between the lift response is shown in Figure 5.12 (e). As the gust amplitude increases, the time taken to respond to the gust also increases slightly, as shown in the figure. The moment response for different gust amplitudes is shown in Figure 5.12 (c). As seen in the case of phase shift, the moment response shows multi-harmonic features. Here the time shift among the responses is predominant, as shown in Figure 5.12 (f), and there is considerable variation in the response amplitudes.

A combined response for the effect of phase shift and gust amplitude for the surge oscillations with respect to the 0-degree case with a 10-degree AoA is shown in Figure 5.13. Curves corresponding to drag, lift, and moment are shown in Figure 5.13 (a), (b), and (c), respectively. As seen from the figure, the time shift between the responses reduces with the gust amplitude.

The responses for heave oscillations for a 10-degree AoA and 90-degree phase shift are shown

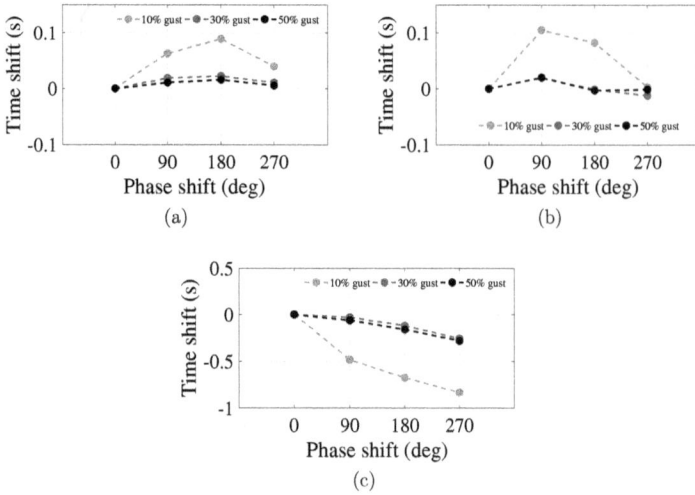

Figure 5.13: Time shift for different gust amplitudes due to surge oscillation: (a) drag, (b) lift and (c) moment.

in Figure 5.14. The drag response, shown in Figure 5.14 (a), shows a considerable increase in the amplitude as the gust amplitude increases. For the lift response, as the gust amplitude increases, the amplitude of the response reduces, as shown in Figure 5.14 (b). Moment response shows an increase in the response amplitude with the increase in gust amplitude, as shown in Figure 5.14 (c). Also, there is a presence of multiple harmonics in the response for higher gust amplitudes. There are phase shifts among the responses for drag, lift, and moment as shown in Figure 5.14 (d), (e), and (f), respectively.

A representative set of results for the pitching oscillation of the aerostat with a 5-degree AoA and 90-degree phase shift between the input signals are shown in Figure 5.15. The drag response, shown in Figure 5.15 (a), has a linear variation in its amplitude and time shift as the wind gust amplitude varies. As the gust amplitude increases, the drag magnitude increases with a slight lag with respect to the 10 % gust case. The variation in the time shift of drag response is shown in Figure 5.15 (d). The lift response, shown in Figure 5.15 (b), also shows an amplitude variation with a predominant difference in the phase among the responses compared to the drag response. The moment response is shown in Figure 5.15 (c), in which the amplitude and phases are identical for the crest of the sinusoidal response, whereas there is a variation in the amplitude and phase for the trough. The phase shift among the lift and moment responses

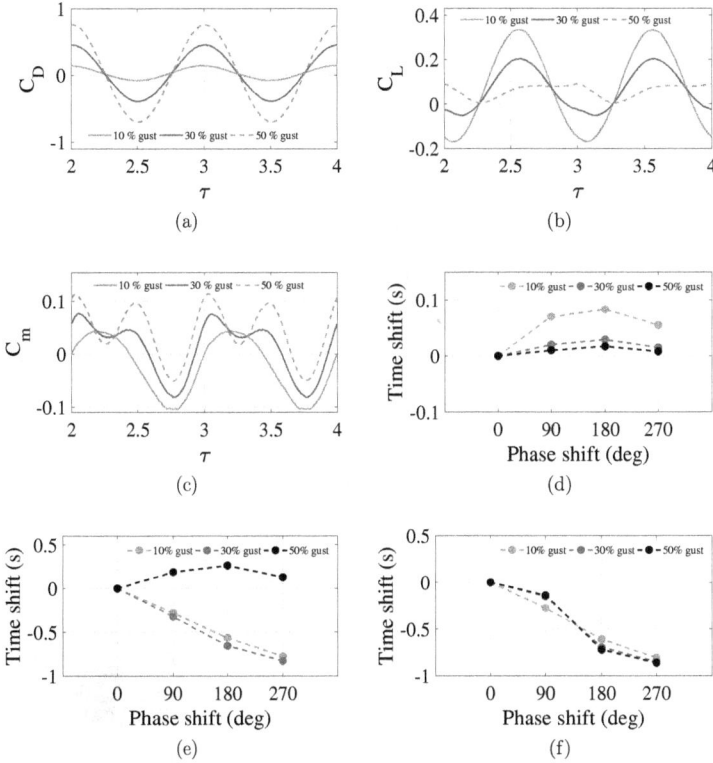

Figure 5.14: Results of the analysis of the effect of gust amplitude for heave oscillation: (a) drag response, (b) lift response, (c) moment response, time shift with respect to the 10 % case for (d) drag, (e) lift, and (f) moment.

is shown in Figure 5.15 (e) and (f).

The variation of the amplitude of the drag, lift, and moment responses of the aerostat with the wind gust amplitude undergoing surge, heave, and pitching oscillations are shown in Figure 5.16 (a), (b), and (c), respectively.

5.3.3   Effect of AoA

For analysing the effect of AoA on the aerodynamic performance of the aerostat, four different AoA were considered: -5, 0, 5, and 10 degrees. The selection of these angles was made so as to capture all the operating conditions of the aerostat.

The responses for the surge oscillation with a 50 % gust amplitude and with a 90-degree

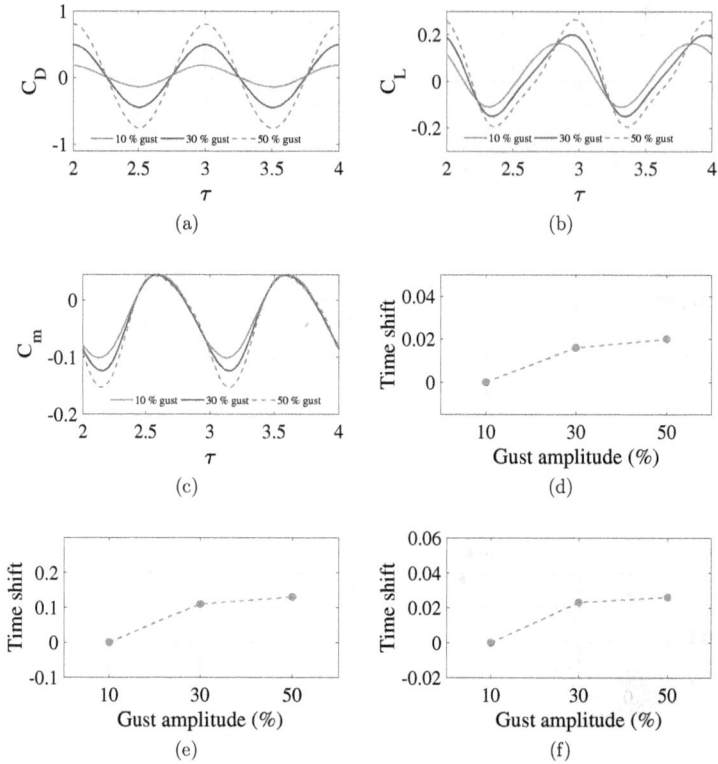

Figure 5.15: Results of the analysis of the effect of gust amplitude for pitch oscillation: (a) drag response, (b) lift response, (c) moment response, time shift with respect to the 10 % case for (d) drag, (e) lift, and (f) moment.

phase shift are shown in Figure 5.17. The drag response, shown in Figure 5.17 (a), shows the responses with almost the same amplitude and phase shift. Figure 5.17 (d) shows the variation of the phase shift with the AoA. The lift response for different AoA is shown in Figure 5.17 (b), in which the 0 AoA case shows negligible variation with an asymmetric response and is plotted on the right-side Y-axis. Here the negative AoA case shows a maximum phase difference from the other cases. As the AoA increases, the lift amplitude increases as well. This is due to the variation in the differential pressure between the top and bottom surfaces of the aerostat. The moment responses are shown in Figure 5.17 (c), in which the maximum amplitude is for the negative AoA. The phase shift among the responses of lift and moment are shown in Figure 5.17 (e) and (f), respectively.

The heave oscillation results for a 50 % gust amplitude and 0-degree phase shift are shown

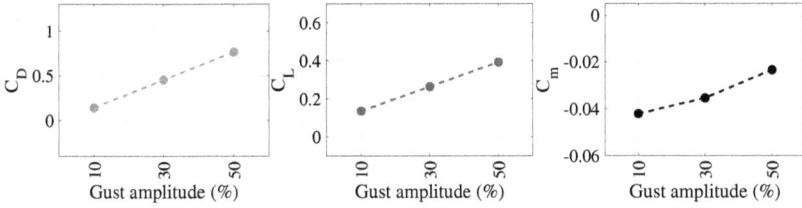

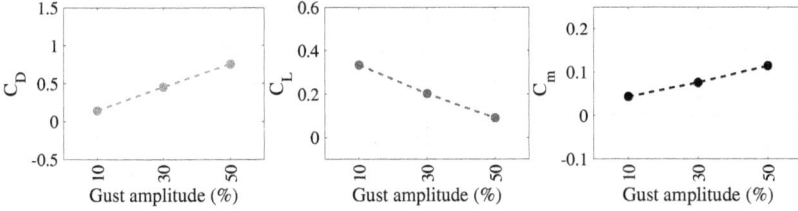

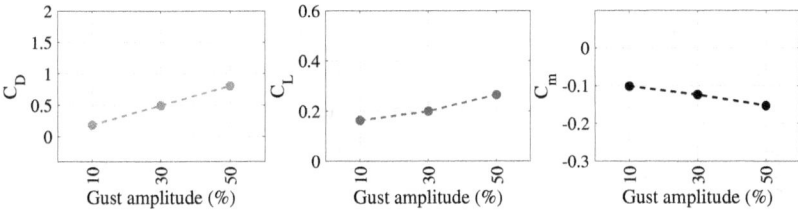

Figure 5.16: Peak amplitudes for different gust amplitudes due to (a) surge, (b) heave, and (c) pitch oscillations.

in Figure 5.18. The drag response remains unchanged as the AoA increases, as shown in Figure 5.18 (a). There is a slight linear variation in the phase shift among the responses, as shown in Figure 5.18 (d). The lift response, shown in Figure 5.18 (b), shows the variation of phase shift with the increase in the AoA. As the AoA reduces from the large positive angle to the negative angle, the phase shift increases linearly, as shown in Figure 5.18 (e). The moment response and the corresponding phase shift results for different AoA are shown in Figure 5.18 (c) and (f), respectively.

The responses of the aerostat undergoing pitch oscillations with 10 % wind gust amplitude and 90-degree phase shift between the input signals are shown in Figure 5.19. The drag response, shown in Figure 5.19 (a), remains almost constant for different AoA, with a slight

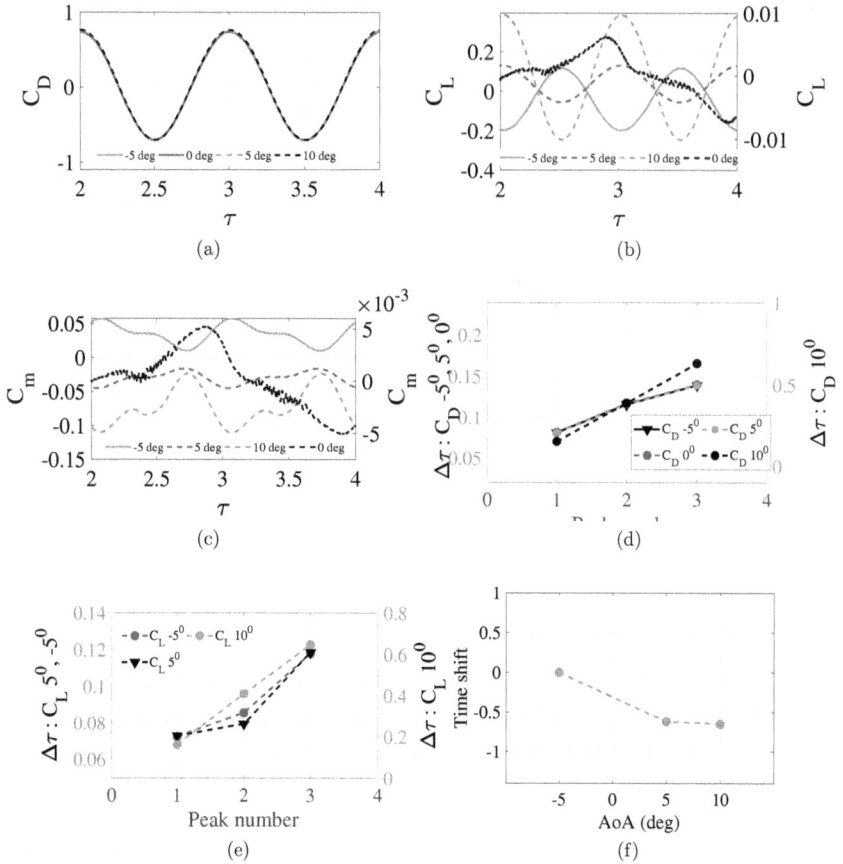

Figure 5.17: Results of the analysis of the effect of AoA for surge oscillation: (a) drag response, (b) lift response, (c) moment response, time shift with respect to the 0 degree case for (d) drag, (e) lift, and (f) moment with respect to the AoA.

variation in the amplitude of the response and phase shift. The -5 degree case is lagging behind the other responses, and there is a linear relationship between the phase difference among the responses, as shown in Figure 5.19 (d). Figure 5.19 (b) shows the lift responses of the aerostat. There is a predominant variation in the amplitude of the responses with the increase in the AoA. The lift amplitude increases as the AoA change from negative to positive. The phase difference between the responses with respect to the -5 degree case is shown in Figure 5.19 (e). The -5 degree case leads all the other three responses, and there is a linear relation for this variation. The moment response of the aerostat is shown in Figure 5.19 (c). The amplitude of the response reduces as the AoA increases. The phase shift of the responses

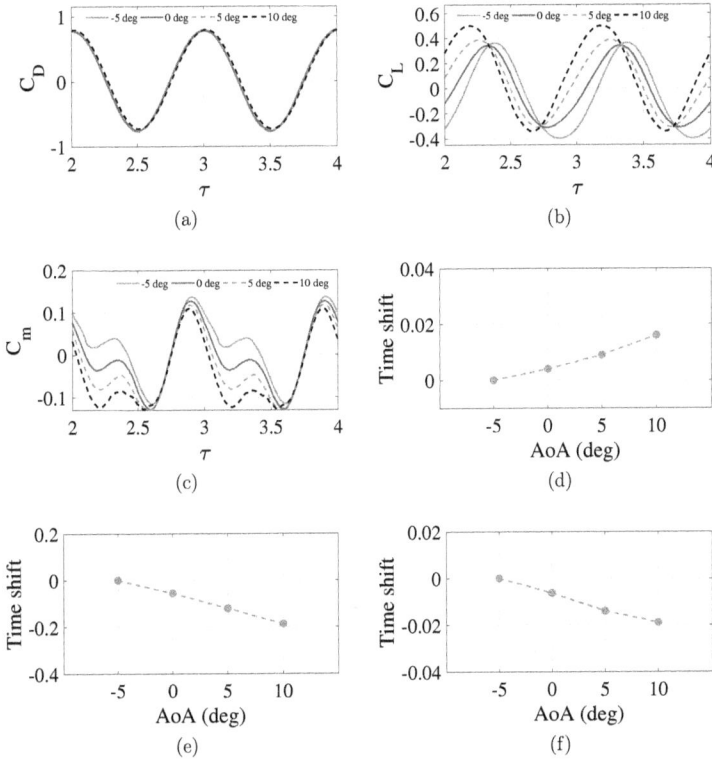

Figure 5.18: Results of the analysis of the effect of AoA for heave oscillation: (a) drag response, (b) lift response, (c) moment response, time shift with respect to the 0 degree case for (d) drag, (e) lift, and (f) moment with respect to the AoA.

with respect to the -5 degree case is shown in Figure 5.19 (f).

The variation of the amplitude of the drag, lift, and moment responses of the aerostat with the AoA undergoing surge, heave, and pitching oscillations are shown in Figure 5.20 (a), (b), and (c), respectively.

5.3.4   Effect of time period

The effect of wind gust duration (time period) on the aerostat was analysed by considering wind gusts with three different time period: 2.7 s, 3 s, and 3.3 s. The analysis was done for an AoA of 10 with a wind gust amplitude of 30 %. Figure 5.21 shows the results of the aerostat undergoing surge, heave, and pitch oscillations. The presented results are for a 0-degree phase

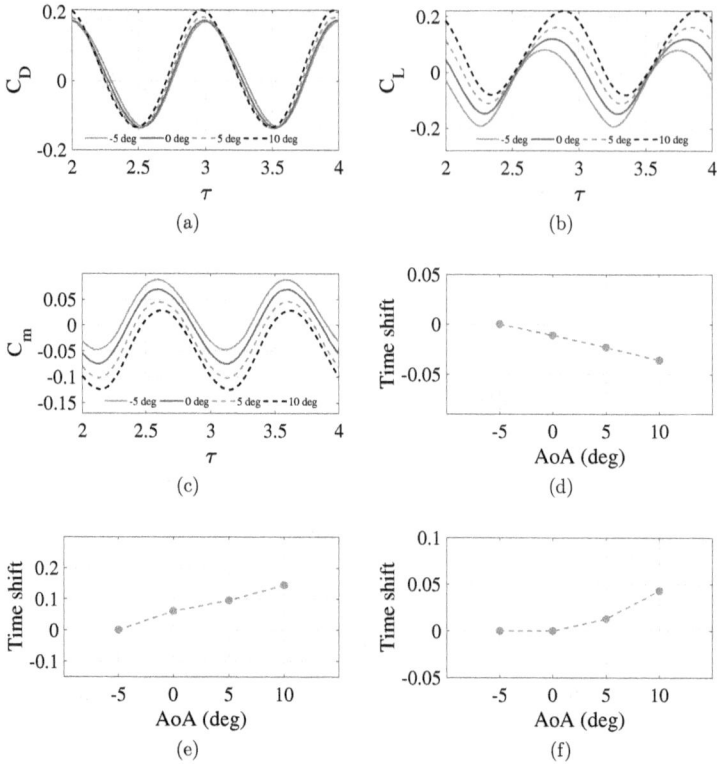

Figure 5.19: Results of the analysis of the effect of AoA for pitch oscillation: (a) drag response, (b) lift response, (c) moment response, time shift with respect to the 0 degree case for (d) drag, (e) lift, and (f) moment with respect to the AoA.

shift condition.

The wind gust duration affects the amplitude of the response rather than the time period or time shift. The forces and moment responses for the surge oscillations are shown in Figure 5.21 (a), (b), and (c), respectively. As the gust length increases, the amplitude of responses decreases. This shows the severity of short duration gusts over their longer counterparts. The drag responses for heave and pitch oscillations shown in Figure 5.21 (d) and g are similar to that of the surge case. The lift response of the heave oscillation case shown in Figure 5.21 (e) shows a large variation in terms of its amplitude. The translational oscillations of the aerostat induce a multi-harmonic component to the moment response, as shown in Figure 5.21 (c) and (f). The additional sinusoidal component increases its effect with the rise in the wind gust length. The forces and moment responses for the pitch oscillations are shown in Figure 5.21

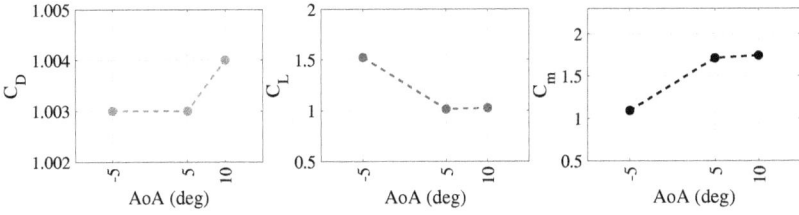

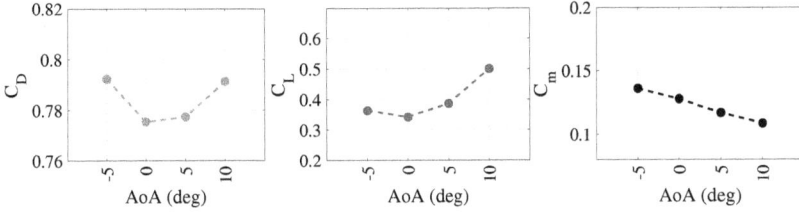

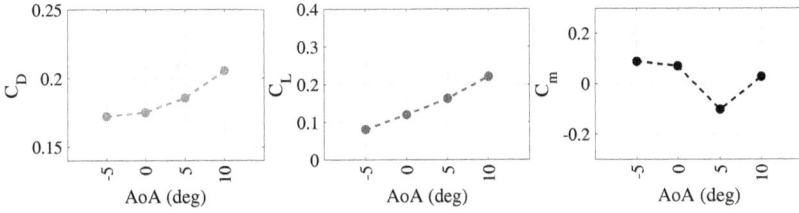

Figure 5.20: Peak amplitudes for different AoA: (a) surge, (b) heave, and (c) pitch oscillations.

(g), (h), and (i), respectively.

The variation of force and moment response amplitudes with the wind gust duration for surge (Figure 5.22 (a)), heave (Figure 5.22 (b)), and pitch oscillations (Figure 5.22 (c)) of the aerostat are obtained. A linear relationship can be drawn between the response amplitude and the wind gust duration. The maximum amplitude variation has occurred for the drag responses and the minimum for the moment responses.

5.3.5 Stability derivatives

The effect of phase shift between the two inputs on the stability derivatives of the aerostat is shown in Figure 5.23. The results shown in the figure are for the -5 degree AoA case with

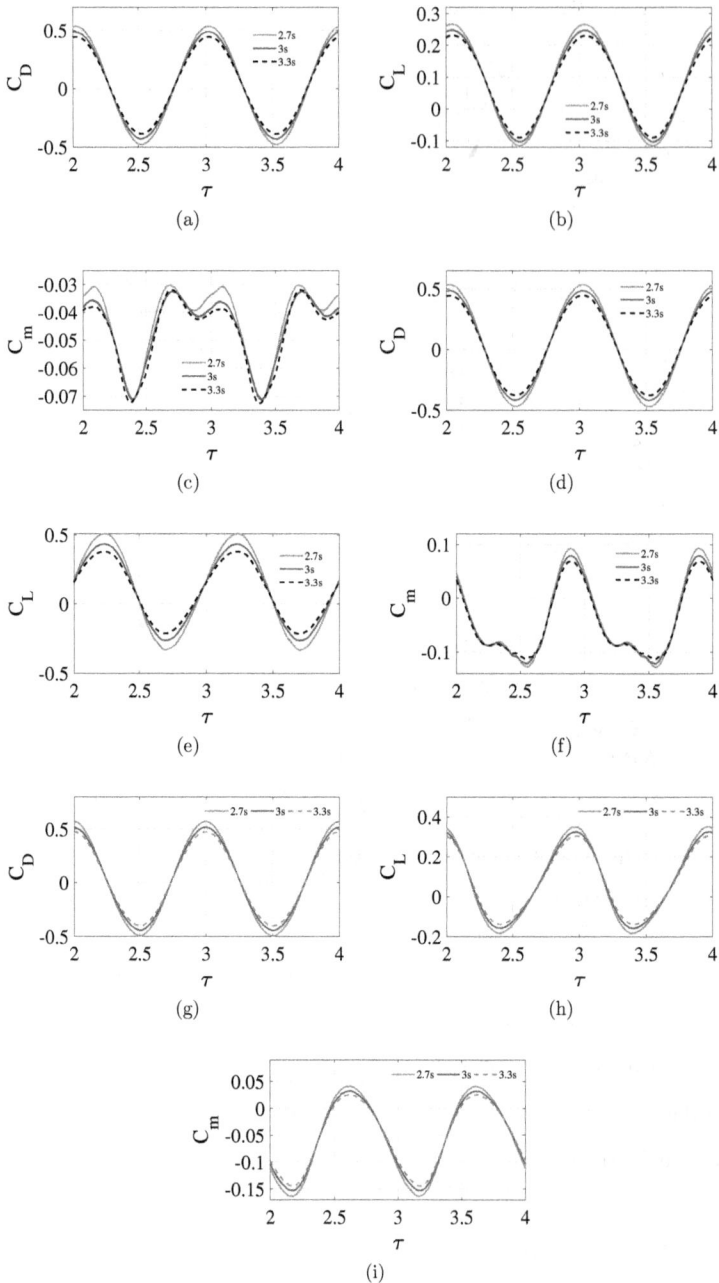

Figure 5.21: Results of the analysis of the effect of the gust time period for surge oscillation (first row), heave oscillation (second row), and pitch oscillations (third row).

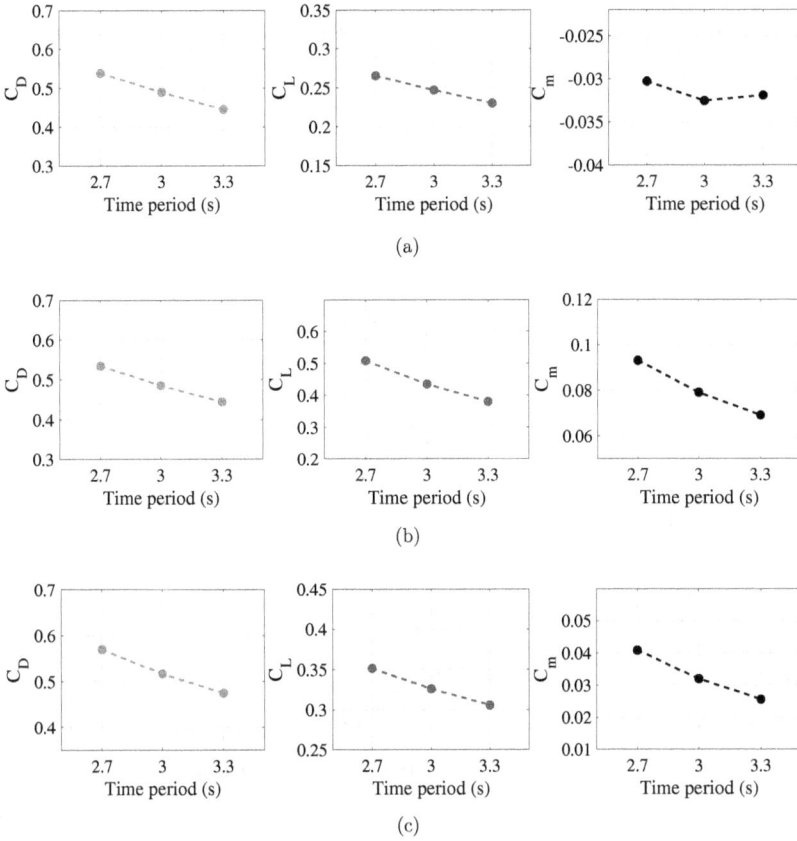

Figure 5.22: Variation of forces and moment amplitude with gust time period for (a) surge, (b) heave, and (c) pitch oscillations.

a 10 % amplitude gust. The drag derivative due to the surge velocity, $\dot{D}_u$, for different phase shifts is shown in Figure 5.23 (a). The magnitude of the drag derivative is minimum for the 90-degree case. This is due to the effective increase in the overall input oscillation. The figure shows that the drag response variation with respect to the surge velocity is minimum for the 90-degree case, and the variation increases with the increase in the phase shift. Figure 5.23 (b) shows the drag derivative due to pitch velocity, $\dot{D}_q$, for different phase shifts. The magnitude of the derivative is comparatively less than the drag derivative due to surge velocity. A phase shift of more than 180 degrees causes the derivative to become negative. The effect of the 90-degree phase shift is reflected in this case as well. From the figure, it is clear that the drag response variation with the pitch velocity decreases with the increase in the phase shift between

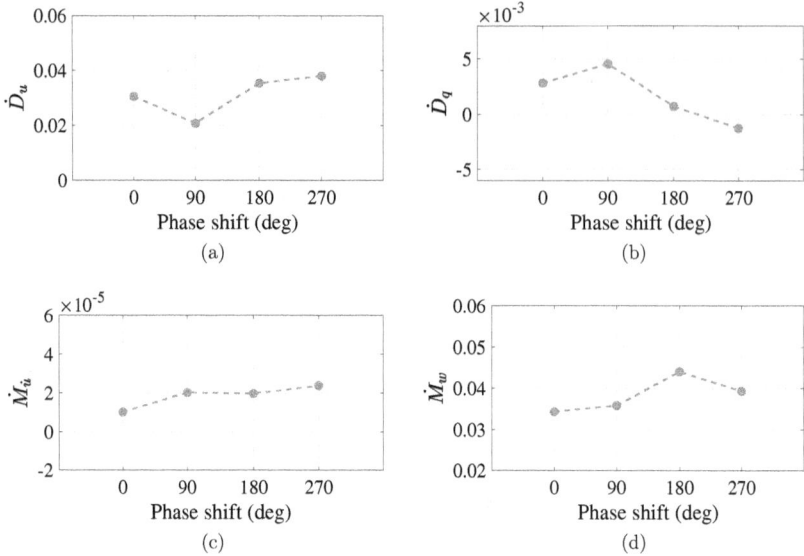

Figure 5.23: Stability derivatives for different phase shifts.

the inputs.

The moment derivative due to the surge acceleration, $\dot{M}_{\dot{u}}$, for different phase shifts between the inputs is shown in Figure 5.23 (c). The variation of moment response due to the variation in the surge acceleration is minimum compared to the lift and drag derivatives. There is a linear relationship between the moment derivative and the input phase shift, as shown in the figure. A slight deviation is visible for the 90-degree case, as discussed previously. Figure 5.23 (d) shows the moment derivative due to heave velocity, $\dot{M}_w$. The magnitude of the derivative increases with the increase in the phase shift between the inputs. There is a slight deviation for the 180-degree case since the case has the maximum phase shift between the peaks of the inputs.

The effect of gust amplitude on the stability derivative is shown in Figure 5.24. The results shown in the figure are for a 0-degree phase shift between the inputs and a 0-degree AoA case. Figure 5.24 (a) shows the drag derivative due to the pitch velocity, $\dot{D}_q$, for three different gust amplitudes. A linear relationship can be seen in the figure. The magnitude of the derivative increases with the increase in the gust amplitude. The drag derivative due to the surge acceleration, $\dot{D}_{\dot{u}}$, for different gust amplitudes is shown in Figure 5.24 (b). Similar to the

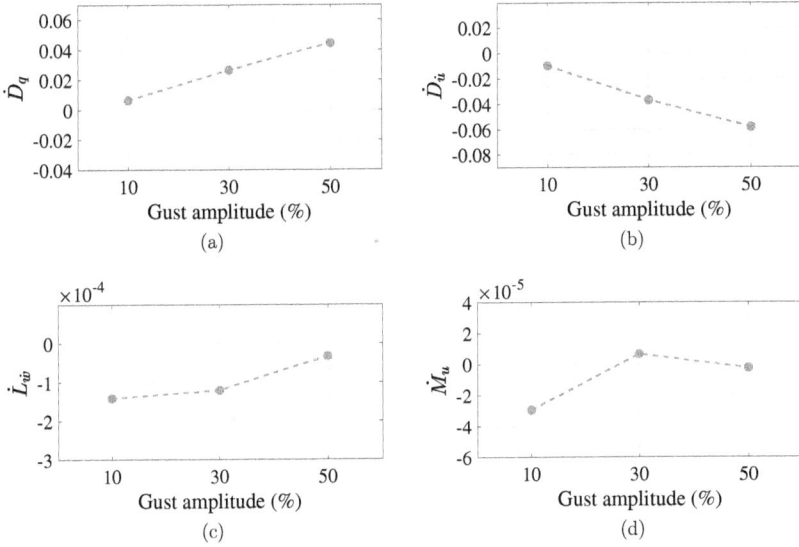

Figure 5.24: Stability derivatives for different gust amplitudes.

previous case, the drag derivative due to surge acceleration shows a linear response with the gust amplitudes. The magnitude of the derivative reduces with the increase in the amplitude of the wind gust. That is, the rate of change of surge velocity decreases the drag response as the wind gust amplitude increases from 10 % to 50 % of the mean wind velocity.

The lift derivative due to the heave acceleration, $\dot{L}_{\dot{w}}$, for different gust amplitudes is shown in Figure 5.24 (c). The magnitude of the derivative is comparatively less and is increasing with the increase in the gust amplitude, as shown in the figure. That is, the rate of change of heave velocity increases the lift responses as the wind gust amplitude increases from 10 % to 50 % of the mean velocity. The moment derivative due to the surge velocity, $\dot{M}_u$, is shown in Figure 5.24 (c) for different gust amplitudes. Similar to the lift derivative, its value is comparatively less. The magnitude of the derivative increases with the increase in the amplitude of wind gusts.

The variation of stability derivatives with respect to the AoA is shown in Figure 5.25. The results shown in the figure are for a 90-degree phase shift and a wind gust amplitude of 30 % case. Figure 5.25 (a) shows the drag stability derivative due to heave velocity, $\dot{D}_w$, for different AoA. As the AoA increases from −5 to 10 degrees, the magnitude of the derivative increases.

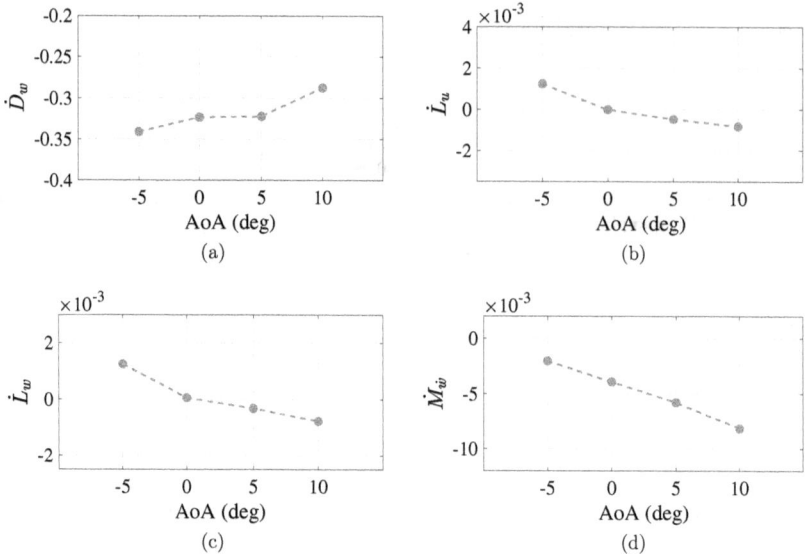

Figure 5.25: Stability derivatives for different AoA.

The negative sign of the derivative term shows that, as the heave velocity increases, the drag response decreases. The lift derivative due to the surge velocity, $\dot{L}_u$, is shown in Figure 5.25 (b). The magnitude of the derivative is decreasing with the increase in the AoA. A zero value is shown for the derivative at the 0-degree AoA. This shows that the lift response remains unchanged with the change in the surge velocity at the 0-degree AoA. The lift derivative decreases its amplitude with the increase in the AoA.

The lift derivative due to heave velocity, $\dot{L}_w$, for different AoA is shown in Figure 5.25 (c). Similar to the lift derivative due to surge velocity, this derivative is decreasing as well with the increase in the AoA. At 0 degree AoA, the derivative becomes zero and has a negative value for the positive AoA and vice versa. Figure 5.25 (d) shows the moment derivative due to the heave acceleration, $\dot{M}_{\dot{w}}$, for different AoA. A linear relationship is seen between the derivative and the AoA. As the AoA increases from -5 to 10 degrees, the derivative value decreases. The negative value of the derivative indicates that the moment response reduces with the increase in the rate of change of heave velocity.

The effect of the wind gust time period on the stability derivative was analysed by doing pitching simulations with three different gust time periods. As explained in Chapter 3, the

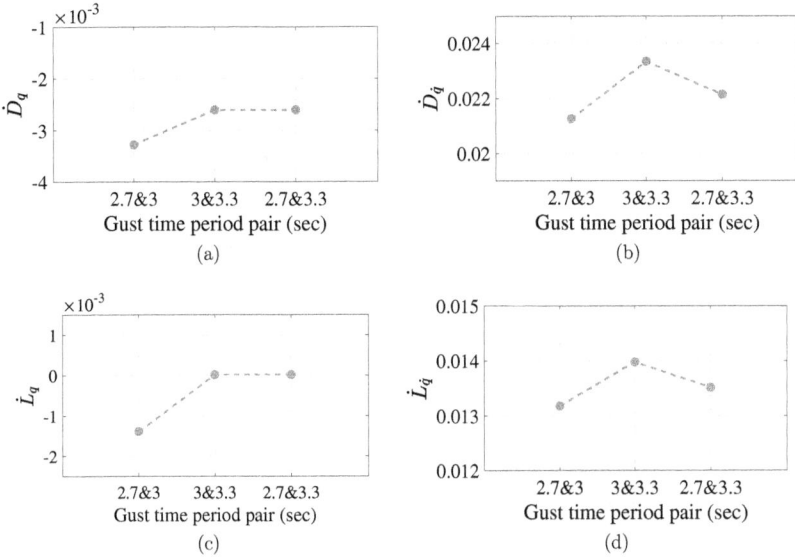

Figure 5.26: Stability derivatives for different gust time period.

estimation of stability derivatives due to pitch oscillation involves two frequencies. These two frequencies were realised by two different time periods of wind gusts. Here in this section, the stability derivatives due to pitch oscillation are presented for three different pairs of gust time periods, namely, 2.7 s & 3 s, 3 s & 3.3 s, and 2.7 s & 3.3 s, as shown in Figure 5.26.

The drag derivative due to pitch velocity, $\dot{D}_q$, is shown in Figure 5.26 (a). The derivative remains the same for the 3 s & 3.3 s and the 2.7 s & 3.3 s pairs because the calculation of the derivative involves the largest time period among the pair, which is 3.3 s for both pairs. A similar response is seen in Figure 5.26 (c), in which the lift derivative due to the pitch velocity remains constant for the second and third pair of the time period. Both the derivatives show a rise in their value with the first two pairs of time periods even though the difference between the time periods remained the same. Figure 5.26 (b) shows the drag derivative due to the pitch acceleration, and Figure 5.26 (d) shows the lift derivative due to the pitch acceleration. Both derivatives show a maximum value for the second pair of the time period for which the sum of the time period is the maximum among the three.

5.4 Summary

The stability derivatives of the aerostat were investigated. A small-amplitude sinusoidal oscillation was used for the simulation of the aerostat motion. A base case aerostat shape was selected for the analysis, and three other aerostat shapes were considered for the comparison of results obtained from the analysis.

The oscillatory motion of the aerostat was simulated using the dynamic mesh motion feature and the zone motion feature available with ANSYS FLUENT. The dynamic mesh motion method was validated using NACA 0012 airfoil test data. The methodology used for the stability derivatives extraction was validated using a prolate spheroid, of which the experimental stability derivatives were available in the open literature.

The static stability derivatives or static aerodynamic coefficients of the four aerostats were presented before the stability derivatives extraction. Then the dynamic stability derivatives of the Zhiyuan aerostat at four different AoA were presented. The results of which were compared with the derivatives of the other three aerostats as well. The unsteady analysis of the aerostat using the combination of unsteady wind and unsteady motion of the aerostat model was also presented. The effect of phase shift between the input wind gust signal and model oscillating signal was analysed. The force and moment responses have shown phase shift compared to the 0 phase shift case.

Stability derivatives, which are used for accurate modelling and stability analysis of aerial vehicles, were investigated for the presented aerostat. By keeping the goal of developing a longitudinal model of the aerostat, the current investigation addressed the extraction of stability derivatives associated with the longitudinal motion of the aerostat.

Chapter 6

Mathematical Model of the Tethered Aerostat

6.1 Introduction

A mathematical model helps in the quantitative analysis of any dynamic system, especially those that have complex components such as mechanical and aerodynamic structures. Analysing the behaviour of the system under different operating conditions, stability analysis, performance evaluation of the system, and control system development for the system are the major applications of a mathematical model. The aim of this chapter is to analyse the performance and behaviour of the single-tethered aerostat. The representation of the dynamics of the aerostat system in the form of a mathematical model will help in the analysis. Mathematical modelling of complex systems such as aerostats plays a vital role in understanding and predicting the unsteady aerodynamic behaviour of that system. The mathematical model for the tethered aerostat system considered in the current research consists of a tether model and an aerostat model.

As a low-altitude operation is considered for the desired application, a simplified tether model is sufficient for the current research. The bending stiffness and damping effect of the tether can be neglected [32]. The aerostat model can be considered a combination of the structural model and the aerodynamic model. The structural model comprises the geometrical attributes of the aerostat, such as weight, buoyancy, tension due to tethering, and so on. Being an aerial vehicle, the aerodynamic part of the model has great importance in aerostat dynamics.

By properly developing an aerodynamic model of the aerostat, the design and analysis of the control system for the vehicle can be implemented in an efficient and effective manner. The mathematical modelling of the aerostat is similar to that of the aircraft in the sense that the dynamics involved in the operation of both vehicles are similar, with some exceptions. The major differences are the buoyancy effect and the added mass effect of the aerostats. These are present for heavier-than-air aerial vehicles but are considered negligible.

This chapter presents the investigation of the mathematical model of the aerostat and the tether. The model parameters (stability derivatives and added mass) of the aerostat are extracted using the methodology explained in Chapter 3. A tether model is also presented and is coupled with the aerostat model to get the complete tethered aerostat system model. A validation study is presented in this chapter for the tether model, and the tether profile for different wind and tether conditions is also reported. The response of the tethered aerostat to different wind conditions is also presented in this chapter.

6.2   Assumptions

It is a fact that the complete dynamics of a physical, real-time system cannot be captured accurately by a mathematical model. It is not required sometimes to capture all the dynamics of a system as well; instead, the interested behaviour alone can be modelled. Assumptions have to be made about the deliberate suppression or neglection of irrelevant details regarding a system before developing its mathematical model. The mathematical model for the aerostat and tether presented in this chapter is based on the following assumptions.

- A longitudinal model of the aerostat is considered. Lateral motion is assumed to be negated by the control surfaces. The forces and the moment considered for the current research are lift, drag, and pitching moment.

- A low-altitude operation is considered. The altitude of the operation is fixed at 100 m above ground level.

- The wind is confined in the axial direction of the aerostat. The components of the wind in the other directions are neglected.

- As the wind is confined to the axial direction of the aerostat, the pitch angle and AoA are assumed to be equal.

- The model is developed for a mean wind velocity of 7.5 m/s, as provided by the NIWE wind data [96], and the model is assumed to perform around this mean velocity.

- The volume of the aerostat is fixed. Changes in the shape of the aerostat and the thermal effect on the envelope are not considered in the current model.

- The stability derivatives estimated using CFD analysis are with respect to the centre of volume as the pitching moment centre.

- A two dimensional model is considered for the tether, as the longitudinal model of the aerostat is considered.

- As the aerostat is assumed to have longitudinal dynamics, the motion of the tether is also assumed to be confined to the vertical plane.

- As the aerostat is not equipped with an active control system, the attitude of operation may vary. Therefore, the length of the tether is allowed to vary with the wind.

- The elongation of the tether due to elasticity is neglected.

- The bending stiffness of the tether is not considered since the length of the tether is small compared to the high-altitude aerostats.

- The leashes (connecting the tether to the aerostat) of the aerostat are assumed to be rigid along with the aerostat. The mass, drag, and damping of the leashes are not accounted for in the current model.

6.3 Tether model

The tether for the aerostat is modelled as a series of segments with equal lengths [32]. Figure 6.1 shows the discretised tether representation with the forces marked. The length of each segment is $d_s = S/n$, where S is the total length of the tether, and n is the number of segments. For n tether segments, there will be $n+1$ nodes. The forces acting on the tether are

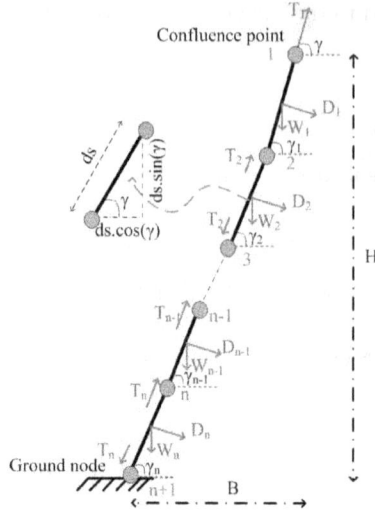

Figure 6.1: Two dimensional discretised tether model.

tension (T), gravitational force (W), and drag force (D). These forces are distributed among the tether segments, as shown in the figure. The altitude of operation is H, i.e., the top node will be at this altitude.

The model of the tether can be obtained by considering the force balance along the horizontal and vertical directions. Forces along the horizontal and vertical directions can be expressed as in Equations (6.1) and (6.2), respectively.

$$T_i \cos \gamma_i + D_i \sin \gamma_i = T_{i+1} \cos \gamma_{i+1} \tag{6.1}$$

$$T_i \sin \gamma_i = D_i \cos \gamma_i + W_i + T_{i+1} \sin \gamma_{i+1} \tag{6.2}$$

where T_i and T_{i+1} are the tension of the ith and $i+1^{th}$ node of the tether; γ_i and γ_{i+1} are the tether angle with the horizontal for the ith and $i+1^{th}$ node of the tether; and D_i and W_i are the normal drag component and weight of the ith segment.

The expression for the angle between the tether segment and the horizontal in terms of the previous node can be obtained as in Equation (6.3), and the expression for tension can be obtained as in (6.4).

$$\gamma_{i+1} = \tan^{-1} \left(\frac{T_i \sin \gamma_i - D_i \cos \gamma_i + W_i}{T_i \cos \gamma_i + D_i \sin \gamma_i} \right) \tag{6.3}$$

116

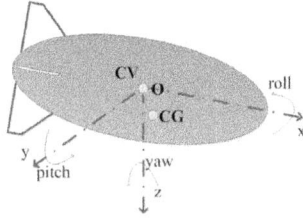

Figure 6.2: Orientation of the aerostat with respect to the body axes.

$$T_{i+1} = \frac{T_i \cos \gamma_i + D_i \sin \gamma_i}{\cos \gamma_{i+1}} \qquad (6.4)$$

The components of tether tension for the model can be obtained by resolving the tension into their respective components.

6.4   Aerostat model

The longitudinal model of the aerostat is developed based on two reference frames, namely the body-fixed reference frame and the inertial reference frame [20]. Figure 6.2 shows the body fixed frame $(Oxyz)$ and its orientation. The origin of the body frame is at centre of volume. Axis Ox points towards the leading edge, Oy points towards the starboard side, and Oz points downwards.

A steady rectilinear operating condition is considered for the model, with V_0 as the free stream wind velocity and Θ_e as the aerostat attitude, as shown in Figure 6.3. The relative velocity of the aerostat with respect to the wind can be expressed as $V_0 = \sqrt{(U_e + V_e + W_e)}$.

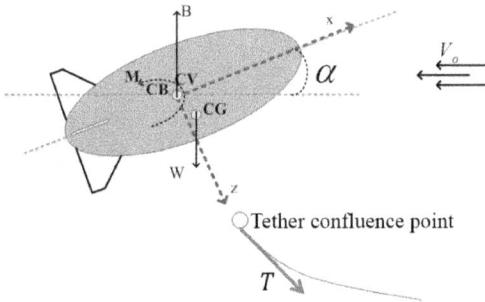

Figure 6.3: Steady rectilinear condition of the aerostat.

For a perturbed condition, the total velocity can be expressed as,

$$\grave{u} = u + U_e$$
$$\grave{v} = v + V_e \qquad (6.5)$$
$$\grave{w} = w + W_e$$

where u, v, and w are the perturbed variables. In steady conditions, these perturbed variables and their derivatives are zero.

The mathematical model of the aerostat consists of Newton's second law of motion applied to each of the concerned DoFs as follows,

$$\mathbf{ma} = \mathbf{F} \qquad (6.6)$$

where $\mathbf{m}$ is the mass, $\mathbf{a}$ is the acceleration, and $\mathbf{F}$ is the disturbing forces. For rotational DoF, the mass, acceleration, and force terms become inertia, angular acceleration, and moment, respectively. The acceleration terms in Equation (6.6) can be obtained by the following calculations. Velocity terms can be expressed as,

$$u = \dot{x} - ry + qz$$
$$v = \dot{y} - pz + rx \qquad (6.7)$$
$$w = \dot{z} - qx + py$$

where x, y, and z are the position coordinates and p, q, and r are the components of angular velocity of the aerostat. Since the aerostat is assumed rigid, $\dot{x} = \dot{y} = \dot{z} = 0$. Thus, Equation (6.7) can be written as,

$$u = -ry + qz$$
$$v = -pz + rx \qquad (6.8)$$
$$w = -qx + py$$

The corresponding acceleration terms can be obtained as similar as Equation (6.7) as follows,

$$a_x = \dot{u} - rv + qw$$
$$a_y = \dot{v} - pw + ru \tag{6.9}$$
$$a_z = \dot{w} - qu + pv$$

The total velocity can be obtained by substituting Equation (6.8) into (6.5) and differentiating it to get the form as in Equation (6.10).

$$\dot{u} = \dot{U} - \dot{r}y + \dot{q}z$$
$$\dot{v} = \dot{V} - \dot{p}z + \dot{r}x \tag{6.10}$$
$$\dot{w} = \dot{W} - \dot{q}x + \dot{p}y$$

Differentiating Equation (6.9) and substituting Equation (6.10) into it to get the acceleration terms as required for Equation (6.6) as follows,

$$\dot{a}_x = \dot{U} - rV + qW - x(q^2 + r^2) + y(pq - \dot{r}) + z(pr + \dot{q})$$
$$\dot{a}_y = \dot{V} - pW + rU + x(pq + \dot{r}) - y(p^2 + r^2) + z(qr - \dot{p}) \tag{6.11}$$
$$\dot{a}_z = \dot{W} - qU + pV + x(pr - \dot{q}) + y(qr + \dot{p}) - z(p^2 + q^2)$$

By using a similar methodology, the angular acceleration terms required for the model can be derived. These acceleration terms can be substituted for 'a' term in the Equation (6.6).

In this work, the longitudinal motion of the aerostat is considered. The lateral motion of the aerostat is assumed to be negated by the control surfaces. The motion variables involved in the decoupled linear longitudinal dynamic model are u, w and q. Setting the lateral dynamic variables and the product terms in Equation (6.6) as zero leads to the linear longitudinal model of the aerostat as given in Equation (6.12).

$$m_x \dot{u} + (ma_z - \dot{D}_{\dot{q}})\dot{q} = D_a + D_T + D_g$$
$$m_z \dot{w} - (ma_x + \dot{L}_{\dot{q}})\dot{q} = L_a + L_T + L_g \tag{6.12}$$
$$J_y \dot{q} + (ma_z - \dot{M}_{\dot{u}})\dot{u} - (ma_x + \dot{M}_{\dot{w}})\dot{w} = M_a + M_T + M_g$$

where, D_a, L_a and M_a are the aerodynamic forces, D_T, L_T and M_T are the forces due to tethers and D_g, L_g and M_g are the forces due to buoyancy and gravitational force. m is the aerostat mass and $m_x = m - \dot{D}_{\dot{u}}$, $m_z = m - \dot{L}_{\dot{w}}$ and $ma_z - m - \dot{M}_{\dot{u}}$ are the added mass terms of the mass matrix.

The aerodynamic force terms can be expressed as the function of stability derivatives of the motion variables involved. Thus, the aerodynamic force terms in Equation (6.12) can be expressed as,

$$D_a = D_e + \dot{D}_u u + \dot{D}_w w + \dot{D}_q q$$

$$L_a = L_e + \dot{L}_u u + \dot{L}_w w + \dot{L}_q q \qquad (6.13)$$

$$M_a = M_e + \dot{M}_u u + \dot{M}_w w + \dot{M}_q q$$

where D_e, L_e, and M_e are the steady state values of the drag, lift, and moment coefficients; and $\dot{\square}_u$, $\dot{\square}_w$, and $\dot{\square}_q$ are the respective dimensional stability derivatives. The added mass coefficients in Equation (6.12) and the stability derivatives required for the aerodynamic part of the model in Equation (6.13) are extracted using CFD-based methodology as explained in Chapter 3. The centre of gravity, centre of buoyancy, and inertia coefficients of the aerostat are obtained from the CAD model analysis.

The buoyancy and gravitational force terms in Equation (6.12) can be expressed as,

$$D_g = (mg - B) \sin \alpha$$

$$L_g = -(mg - B) \cos \alpha \qquad (6.14)$$

$$M_g = -x_b B \cos \alpha - z_b B \sin \alpha$$

where B is the aerostat buoyancy, and x_b and z_b are the coordinates of the aerostat centre of buoyancy.

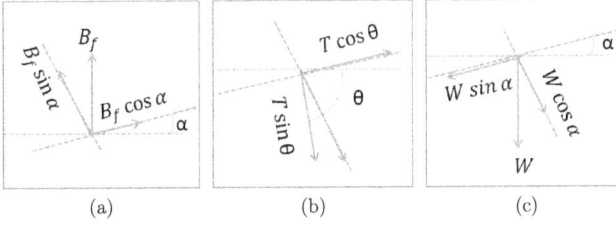

Figure 6.4: Resolving the forces into axial and vertical components: (a) buoyancy force, (b) tether tension, and (c) weight.

Thus, the matrix representation of Equation (6.12) can be given as in Equation (6.15),

$$
\begin{bmatrix}
m - \dot{D}_{\dot{u}} & 0 & ma_z - \dot{D}_{\dot{q}} \\
0 & m - \dot{L}_{\dot{w}} & ma_x + \dot{L}_{\dot{q}} \\
ma_z - \dot{M}_{\dot{u}} & ma_x + \dot{M}_{\dot{w}} & I_y - \dot{M}_{\dot{q}}
\end{bmatrix}
\begin{bmatrix}
\dot{u} \\
\dot{w} \\
\dot{q}
\end{bmatrix}
=
\begin{bmatrix}
D_e \\
L_e \\
M_e
\end{bmatrix}
+
$$

$$
\begin{bmatrix}
\dot{D}_u & \dot{D}_w & \dot{D}_q \\
\dot{L}_u & \dot{L}_w & \dot{L}_q \\
\dot{M}_u & \dot{M}_w & \dot{M}_q - ma_x U_e - ma_z W_e
\end{bmatrix}
\begin{bmatrix}
u \\
w \\
q
\end{bmatrix}
+
\begin{bmatrix}
D_T \\
L_T \\
M_T
\end{bmatrix}
+
\begin{bmatrix}
(mg - B)\sin\alpha \\
-(mg - B)\cos\alpha \\
-x_b B \cos\alpha - z_b B
\end{bmatrix}
\tag{6.15}
$$

The terms in the matrix, $\begin{bmatrix} D_T & L_T & M_T \end{bmatrix}^T$, of Equation (6.15) represents the axial and vertical forces and the moment due to the tether. As the drag due to the tether segments is comparatively negligible compared with the aerostat drag, the tension forces alone are considered for the simulation. Resolving the tension of the tether will give the D_T, L_T, and M_T. The tether force components can be obtained based on the formulations presented in Section 6.3, and the forces are resolved as per Figure 6.4.

6.5 Tethered aerostat model

The tether model and the aerostat model are coupled together by matching the geometrical constraints of the models. The attitude of the top node of the tether is provided as the confluence point of the aerostat. The resolved horizontal and vertical forces of the aerostat are applied to the top node of the tether.

A flow chart explaining the complete scheme for the mathematical modelling of the tethered

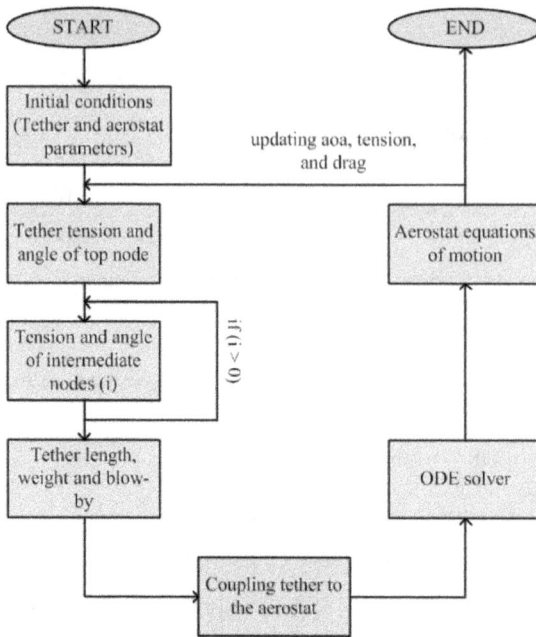

Figure 6.5: Simulation scheme adopted for the tethered aerostat simulation.

aerostat is shown in Figure 6.5.

MATLAB 2020b is used for the simulation of the tethered aerostat model. The simulation is started by providing the required geometrical parameters of the tether and aerostat. As the stability derivatives of the aerostat are obtained for a mean wind speed of 7.5 m/s, the tethered aerostat model is also simulated for the same velocity.

The simulation starts by calculating the tether tension and the tether angle of the top node of the tether by considering the steady-state conditions, and the equilibrium of forces can be given as specified in Equation (6.16).

$$T \sin \theta = B - W$$
$$T \cos \theta = D$$

(6.16)

The tension and angle of the following segments of the tether can be calculated using the equations specified in Section 6.3 in a recursive manner. As the tether segments are connected in series, this calculation can continue till the nth node is reached. Once the full tether profile is obtained, the total length of the tether, the weight of the tether, and the blow-by caused

122

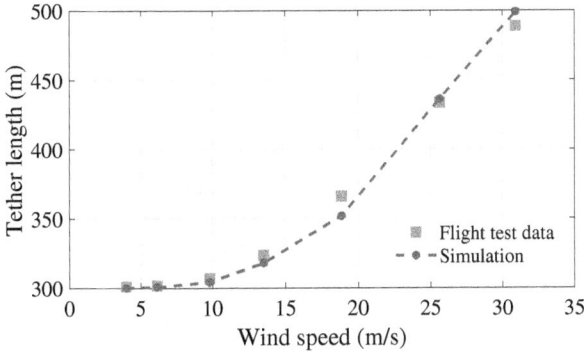

Figure 6.6: Validation of the presented tether model.

by the wind can be calculated from the X and Z coordinates of the tether segments. This procedure can be used to obtain the tether profile for various wind conditions. As the position coordinates of the top node of the tether can be obtained from the tether profile, the position coordinates of the aerostat need not be transformed to the earth-fixed reference frame. They can be directly obtained from the equations of motion of the aerostat and tether.

The ordinary differential equations of the aerostat model were solved using the ODE45 numerical solver available with MATLAB 2020b. As the drag calculation of the aerostat is required for the tether model, the complete tethered aerostat model equations are considered in the ODE45 function. Thus, the drag and, thereby, the tether profile are updated for each iteration of the solver.

The coupling of the tether model to the aerostat model is done by providing the tension force from the tether to the aerostat and by specifying the initial position of the aerostat based on the top node position of the tether.

6.6   Results and discussion

The simulation results of the tether and aerostat models are presented in this section. MAT-LAB 2020b is used for the analysis. The model parameters of the tethered aerostat used in this research are shown in Table 6.1. The stability derivatives used for the model are dimensional values, whereas the stability derivatives presented in Chapter 5 are non-dimensional values.

The tether model presented in Section 6.3 is validated using the flight test data of a 250 m^3

Table 6.1: Aerostat and tether model parameters used for the simulation analysis.

Parameter	Description	Value
W	Weight of the tethered aerostat	1129 kg
B	Buoyancy	1315 kg
a_x	Aerostat centre of gravity - X coordinate	0.4 m
a_z	Aerostat centre of gravity - Z coordinate	0 m
x_b	Aerostat centre of volume - X coordinate	0 m
z_b	Aerostat centre of volume - Z coordinate	0 m
I_y	Moment of inertia of aerostat envelope about Y axis	219 kgm^2
U_e	Wind velocity along the axial direction	7.38 m/s
W_e	Wind velocity along the vertical direction	1.30 m/s
$\dot{M}_w$	Moment derivative due to vertical velocity	-11.3229
$\dot{M}_q$	Moment derivative due to pitch velocity	-0.6369
$\dot{D}_{\dot{u}}$	Drag derivative due to axial acceleration	-16.63477
$\dot{D}_{\dot{q}}$	Drag derivative due to pitch acceleration	0.8501547
$\dot{L}_{\dot{w}}$	Lift derivative due to vertical acceleration	-115.9107
$\dot{L}_{\dot{q}}$	Lift derivative due to pitch acceleration	-4.476338
$\dot{M}_{\dot{u}}$	Moment derivative due to axial acceleration	-7.444886
$\dot{M}_{\dot{w}}$	Moment derivative due to vertical acceleration	-160.5588
$\dot{M}_{\dot{q}}$	Moment derivative due to pitch acceleration	-14.21907
$\dot{D}_u$	Drag derivative due to axial velocity	-16.9854
$\dot{D}_w$	Drag derivative due to vertical velocity	-15.3459
$\dot{D}_q$	Drag derivative due to pitch velocity	-6.96121
$\dot{L}_u$	Lift derivative due to vertical velocity	-0.65086
$\dot{L}_w$	Lift derivative due to vertical velocity	-95.4811
$\dot{L}_q$	Lift derivative due to pitch velocity	1.305291
$\dot{M}_u$	Moment derivative due to axial velocity	-59.257
m_T	Tether mass per unit length	0.3 kg/m
d_s	Length of tether segment	0.1 m
Cd_T	Tether drag coefficient	1.17

aerostat [32]. The variation of the tether length for different wind amplitudes was compared, as shown in Figure 6.6. The results show considerable similarity with the reported results and thus validate the presented tether model.

The tether model considered in this research constitutes a number of discrete segments, as shown in Figure 6.1. The discretisation of the tether considered for this research is done after a set of studies that investigated the effect of the number of segments and the length of each segment on the tether profile. Different tether segment lengths were considered for the model, and the response of the tether profile was simulated for a wind with an amplitude of 12 m/s, as shown in Figure 6.7 (a). It can be seen that for the longer segments, the model

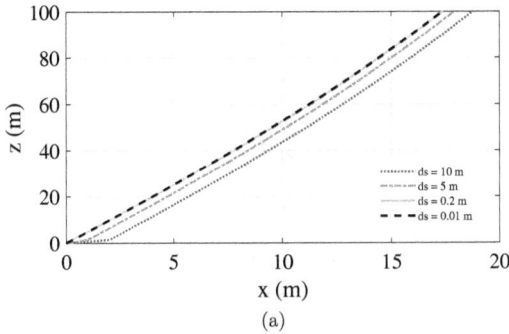

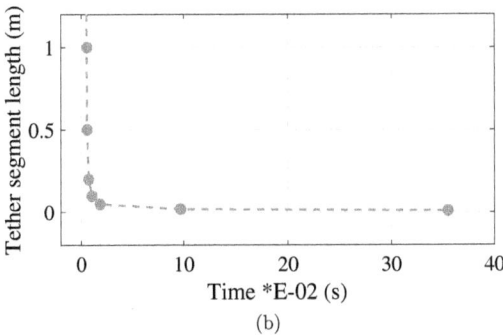

Figure 6.7: a) Tether profile obtained for different tether segment lengths and (b) Computational time variation with the tether segment length.

failed to demonstrate the tether profile accurately. As the segment length reduces to 1 m, the tether profile converges. Further reduction in the segment length is not showing any further improvement. As the length of the tether segment reduces, the number of tether segments increases, which adds to the computational cost. A comparison of the simulation time for different segment lengths is shown in Figure 6.7 (b). The computational cost increases as the number of segments increases. By considering a trade-off between accuracy and computational cost, the tether segment length is selected as 0.1 m for further analysis.

Even though the simulation studies reported in the upcoming sections of this chapter deal with fixed wind conditions, the ability of the tether model to respond to different wind conditions has to be investigated. Here in this section, the tether profile was simulated for various wind amplitudes, and the results are shown in Figures 6.8 and 6.9. Two conditions were considered: (i) the altitude of operation was kept constant by allowing the tether to vary its length, and (ii) the length of tether was kept constant. Figure 6.8 (a) shows the response of the tether

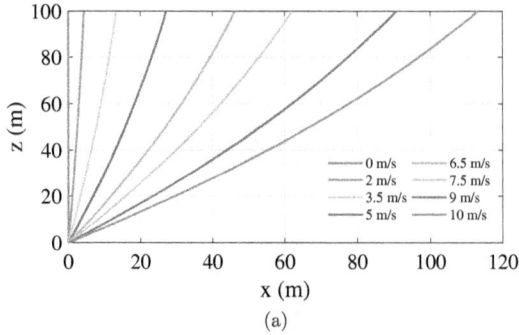

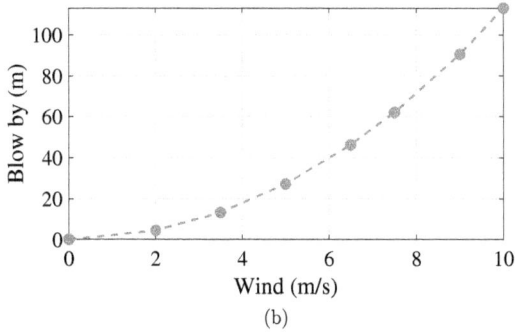

Figure 6.8: (a) Tether profile for different wind amplitudes with fixed altitude of operation and (b) blow-by with a fixed altitude of operation.

to the first condition. This condition demands active control over the tether length based on the altitude of the top node and the tension on the tether, which is beyond the scope of this thesis. The tether should be released if the tension grows beyond a particular limit and the altitude is reducing, and the tether should be retracted if the tension is low and the altitude is increasing. Figure 6.9 (a) shows the response of the tether to the second condition. Since the length of the tether is fixed, the altitude of operation cannot be maintained at 100 m. This condition demands active control of the aerostat to maintain its altitude by opposing the motion due to wind, which is beyond the scope of this thesis. It will make the control system complex, and the payload capacity will decrease due to the additional control equipment on board. Figure 6.8 (b) and Figure 6.9 (b) show the axial (blow-by) and vertical displacement of the top node of the tether for a fixed amplitude of operation and a fixed tether length operation. For the previous set of analyses, fixed weight, buoyancy, lift, and drag forces were considered to replace the aerostat forces. The altitude of the top node of the tether, the blow-by of the

126

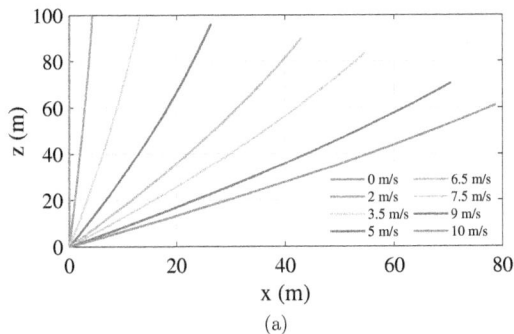

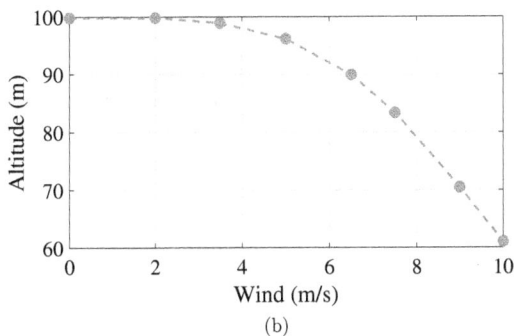

Figure 6.9: (a) Tether profile for different wind amplitudes with fixed tether length and (b) dip in the altitude with a fixed length tether.

tether due to wind, and the total tether length for each wind condition may vary with the actual aerostat model incorporated into the tether model.

The results of the simulation analysis done on the single-tethered, low-altitude aerostat are presented here. As discussed in Chapter 5, the stability derivatives play a significant role in the aerodynamic model of the aerostat. The CFD-based methodology presented in Chapter 3 was used to extract the stability derivatives of the aerostat at a mean wind velocity of 7.5 m/s. The process involved small amplitude sinusoidal oscillations about the mean wind velocity, which allowed the model to work around that wind velocity.

The response of the tethered aerostat model for the mean wind velocity of 7.5 m/s is shown in Figure 6.10. The aerostat came to rest after 500 s, as shown in the figure. The forward velocity of the aerostat settles down to zero after a peak overshoot. The vertical velocity also reaches zero after a negative overshoot, as shown in the figure. The presented model is able to dampen out the oscillations and allows the model to settle down in a stable condition, as

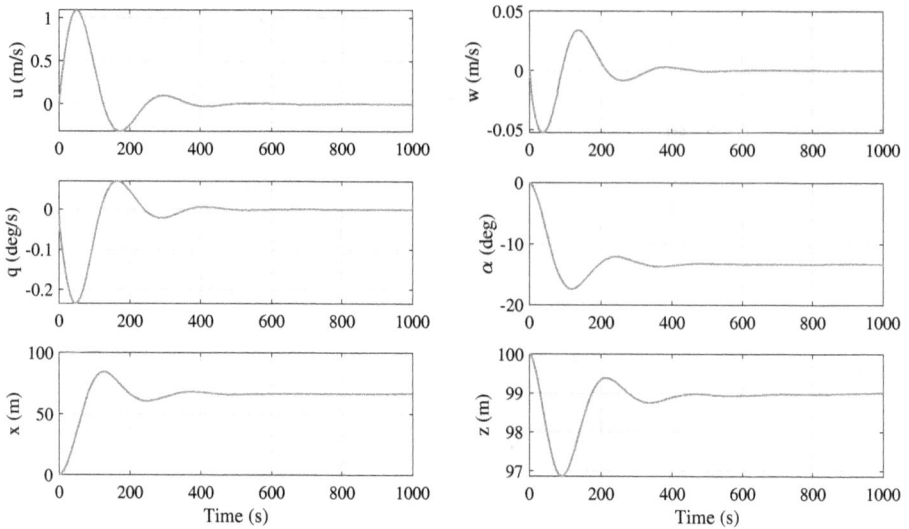

Figure 6.10: Response of the single-tethered aerostat undergoing a steady wind of 7.5 m/s.

shown in the figure.

The pitch angle in the response is found to be negative, as shown in Figure 6.10. This can be explained as the effect of the selection of the pitching centre for the aerostat while estimating the stability derivatives. The aerostat centre of volume was used as the pitching centre for the current analysis, as shown in Figure 6.11 (a). As the leashes of the aerostat were assumed to be rigid along with the tether, the pitching of the aerostat has to be about the confluence point of the tether, as shown in Figure 6.11 (b). This deviation in the pitching centre caused improper pitching of the aerostat and drove the aerostat to a negative pitch angle, as shown

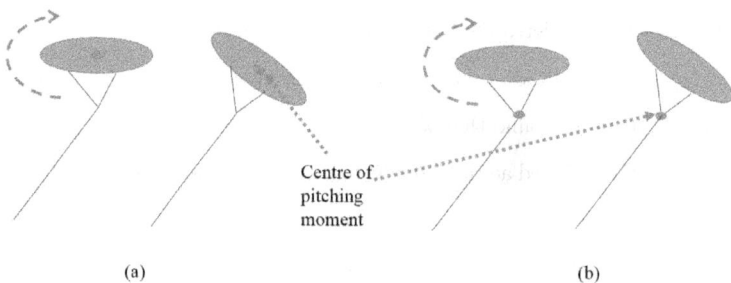

Centre of pitching moment

(a)   (b)

Figure 6.11: Schematic diagram showing the pitch centre of tethered aerostat: (a) aerostat centre of volume as pitch centre and (b) tether confluence point as pitch centre.

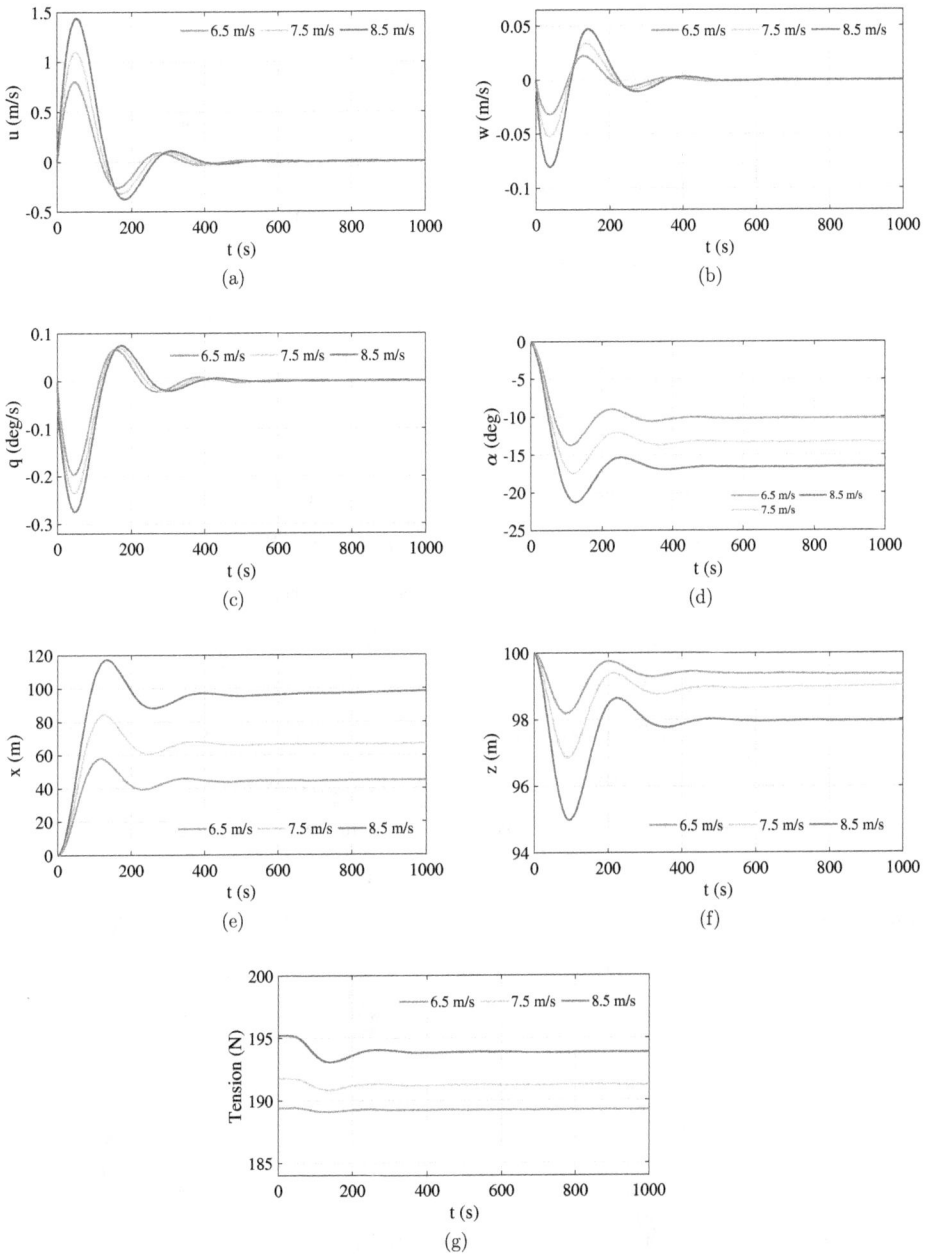

Figure 6.12: Response of the tethered aerostat for wind velocities close to the operating velocity: (a) forward velocity, (b) vertical velocity, (c) pitch rate, (d) pitch angle, (e) horizontal position, (f) vertical position, and (g) tether tension.

in Figure 6.10.

The position of the aerostat can be obtained from the X and Z coordinates of the aerostat, as shown in Figure 6.10. As the equations of motion were initiated with the position coordinates of the topmost node of the tether, the aerostat position can be obtained directly without any axis transformation. The altitude of operation was found to be slightly less than 100 m which shows the necessity of an active control system for the aerostat. In a similar manner, the blow-by caused by the wind drifted the aerostat up to 68 m from its initial position. An active control surface will help in controlling the blow-by of the aerostat.

As mentioned previously, the presented model works in the vicinity of the mean wind velocity of 7.5 m/s. The responses of the tethered aerostat model for three different wind velocities are shown in Figure 6.12.

As the wind velocity increases, the forward velocity of the aerostat increases with a hike in the overshoot, as shown in Figure 6.12 (a). The time taken for the settling of the response also increases with the wind. Similar behaviour is exhibited by the vertical velocity, as shown in Figure 6.12 (b). The pitch angle shows a considerable variation with the wind velocity, as shown in Figure 6.12 (d). This behaviour once again suggests the need for a proper pitch control system for the aerostat. The negative pitch was the result of the selection of the aerostat centre of volume as the pitching moment centre for the estimation of the stability derivatives.

The tension force on the aerostat tether is shown in Figure 6.12 (g). The magnitude of tension is lowest for the 6.5 m/s wind, and the maximum tension was obtained for 8.5 m/s wind, which is the maximum wind velocity considered. As the wind velocity increases, the amplitude of tension force on the tether shows a large initial value and then settles down to a lower amplitude. For lower wind velocities, this behaviour is not dominant. The time taken to settle down the tension force response is also increasing with the increase in wind velocity.

6.7 Summary

An investigation of the longitudinal dynamic model of the single-tethered aerostat was carried out in this chapter. The model parameters (stability derivatives and added mass terms) of the aerostat aerodynamic model estimated from the CFD-based analysis presented in Chapter 5 were investigated.

A two-dimensional model for the aerostat tether was developed. The motion of the tether was confined in the vertical plane of symmetry. The dynamics of the tether were investigated by discretising the entire tether length into n segments. The forces considered for the model were tether tension, drag, and weight. The effects of elongation and bending of the tether were neglected. The presented model was validated using experimental data available for a tethered aerostat flight test. Tether length for different wind conditions was compared with the current simulation results for validation. The effect of segment length on the tether profile was analysed, and it was found that as the tether length decreases (the number of tether segments increases), the tether profile converges to a smooth profile. A shorter segment length helps in the proper discretisation of the tether at the expense of computational cost. A trade-off between the number of segments and the computational cost was considered since a further reduction in the tether segment length was not changing the tether profile much. For further analysis, a segment length of 0.1 m was selected. The tether profile for different tether configurations and wind conditions was investigated. The leashes connecting the tether to the aerostat were assumed to be rigid along with the aerostat.

A longitudinal decoupled dynamic model of the aerostat was considered in the current study by assuming that the lateral dynamics were negated. The added mass terms and the stability derivatives of the aerostat were estimated using a CFD-based analysis. The model was found to be valid around the mean wind velocity with which the stability derivatives were estimated. The tether model was coupled to the aerostat model by initiating the aerostat with the tether tension, weight, and position of the top node of the tether. The MATLAB ODE45 numerical solver was used for solving the aerostat equations of motion along with the tether dynamics. The model was able to respond to different wind conditions as expected. The negative pitching of the aerostat was due to the fact that the stability derivatives were investigated by considering the aerostat centre of volume as the pitching moment centre instead of the tether confluence point.

Chapter 7

Conclusion and Scope for Further Applications

7.1 Introduction

Wind data plays a significant role in the wind energy harvesting sector. The conventional method for collecting short- and long-term wind data using a mast or tower suffers from a number of drawbacks, such as its inability to vary the altitude of operation, the requirement for a permanent structure to be installed, the inability to relocate it after its usage, the limited altitude of operation, and so on. This thesis proposes a single-tethered aerostat for wind data collection at lower altitudes. Its ability to operate at varying altitudes and its ability to dismantle and transport it to different locations to reuse it for other applications make it a superior option over the conventional mast-based method.

The major contribution of this thesis is the aerodynamic analysis of the aerostat and the mathematical model of the aerostat based on the stability derivatives. The performance of the aerostat considered for the research has been investigated for various wind and operating conditions. Wind gust models of different velocity profiles, amplitudes, time period, AoA, and aerostats with different shapes were considered for the investigation. A scaling methodology for the aerostat responses was also proposed for the presented results.

For analysing the operating conditions and investigating the performance of the tethered aerostat, a mathematical model suitable for the considered operating conditions was necessary. In this thesis, a longitudinal dynamic model of the tethered aerostat was considered. A 2D model was used for the tether, and the mathematical model of the aerostat was developed based on the stability derivatives. A CFD-based methodology was used to extract the stability derivatives of the aerostat.

A brief summary and major observations of the thesis are presented in this chapter. As every research project paves the way for new research objectives or scopes, some of the identified future scopes of the current research are also discussed in this chapter.

7.2 Summary of the research

A low-altitude single-tethered aerostat was proposed for wind measurement applications that need to be analysed under unsteady wind conditions. This research began by investigating the effect of unsteady wind gusts on the aerodynamic characteristics of an aerostat. A 3D model of the four-winged aerostat was modelled for the analysis. The obtained model was validated using experimental data from the literature. The unsteady conditions in the aerostat operating region were included as wind gusts in the simulation. Wind data from the Gujarat coastal area was used for the gust modelling. ANSYS Fluent was used for the simulation analysis.

The effects of AoA, gust amplitude, gust duration, and gust length on the aerodynamic performance of the aerostat were investigated using different gust conditions. It was found that the variation in the AoA was not affecting the aerostat drag very much. The minimum drag obtained was for the zero degree AoA. It should be noted that at zero AoA, the lift is also minimal. So, a positive AoA is preferable for the aerostat for its stable operation. The peculiar variation in the contribution of the pressure effect and the viscous effect to the total drag causes a phase shift in the force response with respect to the gust profile. The phase shift increases with the progressing peaks in the response. The phase shift in the moment response is comparatively negligible, and the moment magnitude decreases for the progressing peaks. In the case of gust amplitude variation, a noticeable variation in the force and moment responses was observed. The growth in the magnitude of the peaks in the force responses was linear with the gust amplitude, whereas the central peak of the moment response alone shows a linear growth.

The effect of gust length was investigated using a wide range of time periods. It was found that as the gust length increased, the impact it had on the aerostat's performance decreased.

The overshoots in the force responses diminish as the gust length increases. A comparatively smooth response was obtained for drag and lift forces. On the other hand, a slight increase in overshoots was observed for the moment response. The amplitude of the peaks of the force response showed a linear decrease for the small duration, and the slope reduced as the duration went higher. The moment response peaks had a comparatively constant amplitude throughout the range. A phase shift in the force and moment responses was observed, with a linear variation for the drag response. For the lift response, the variation in phase shift showed a slightly non-linear variation, and for the moment response, it showed a highly non-linear variation. Three different aerostat geometries were considered for the analysis of the effect of aerostat shape. The drag and lift characteristics of the geometry showed an improvement with the reduction in diameter. The characteristics were not desirable for a shape with a larger diameter. Among the different gust shapes considered for the analysis, the symmetric gusts generated symmetric force and moment responses. Among the symmetric profiles, the cosine gust had the maximum amplitude, but the response for it was smoother among all the profiles. It was due to the slow rising and falling of the gust. As the symmetry in the gust profile diminished, the response demonstrated overshoots. Gusts with a very short rising time had a more severe effect on the response than gusts with a short falling time.

From the simulation studies conducted on the aerostat model with different wind gusts, a common trend was observed among the responses. An investigation on the scaling of the aerostat responses for different AoA, gust amplitudes, aerostat shapes, and gust durations was conducted. A scaling methodology was presented in which the gust responses were scaled to represent them as a single curve. A fifth-order polynomial representation of the response curves was used for the scaling. This scaling methodology can be extended to predict the responses of gust conditions that were not considered for the simulation.

Stability derivatives, which are used for accurate modelling and stability analysis of aerial vehicles, were investigated for the presented aerostat. By keeping the goal of developing a longitudinal model of the aerostat, the extraction of stability derivatives associated with the longitudinal motion of the aerostat was carried out. The CFD-based extraction methodology was extended for the aerostat application. A set of validation studies were conducted before proceeding with the aerostat model. The results of the validation study were promising, which gave confidence to extend the methodology for the extraction of the stability derivatives of the presented aerostat. A study on the effect of gust amplitude, gust time period, AoA, and phase shift between inputs on the stability derivatives was presented.

A comparison of the aerodynamic performance and stability derivatives of four different aerostats (Zhiyuan, GNVR, HAA, and NPL) was conducted. The static stability derivatives, or static aerodynamic coefficients, of the four aerostats were presented before the stability derivative extraction. The Zhiyuan shape showed the maximum drag among the four shapes because of its large surface area and large front area. Being the shape with the least surface area, the NPL shape possesses the least drag. The lift characteristics of the Zhiyuan shape were better than all the other shapes due to its large surface area, which contributed to the differential pressure distribution. The lift drag ratio of the aerostats showed that the Zhiyuan aerostat has the best static aerodynamic characteristics of the four aerostats.

The stability analysis and the control system design for the aerostat can be improved by considering the dynamic stability derivatives involving the small-amplitude motions of the aerostat. The dynamic stability derivatives of the Zhiyuan aerostat at four different AoAs were presented. The results of which were compared with the derivatives of the other three aerostats as well. The drag and lift performance of the Zhiyuan aerostat is superior among the four aerostats. This will help the aerostat be controlled with less effort during the station keeping and attitude control tasks. The NPL aerostat had better moment stability for pitching acceleration. This will make the Zhiyuan aerostat demand more control effort for the pitching manoeuvre.

Among the four aerostats considered, the Zhiyuan aerostat was suited for the low-altitude wind measurement application. The aerodynamic geometry of the Zhiyuan aerostat compared to the other three aerostats was the main reason for this peculiar performance. The static lift-drag performance of the Zhiyuan aerostat is better among the four aerostats presented. The dynamic stability analysed using the stability derivatives also suggests the superiority of the Zhiyuan aerostat among the four aerostats. The drag and lift performance of the Zhiyuan aerostat is such that it aids station keeping and attitude control efforts. Even though the moment stability is slightly below the NPL aerostat, the Zhiyuan aerostat can be operated for the pitching manoeuvre with some added control effort.

An unsteady analysis of the aerostat using the combination of unsteady wind and unsteady motion in the aerostat model was investigated. Sinusoidal wind gusts and sinusoidal translational oscillations of the model, along with the axial (surge) and vertical (heave) directions and rotational oscillations about the pitch axis, were considered. The study included the effects of phase shift between the wind and the oscillation sinusoidal signals, the effect of the wind gust amplitude, the effect of the AoA, and the effect of the time duration of the wind gust. A validation study of the selected aerostat model for the current study was done using experimental test data. The grid motion feature used for the pitching oscillation and the dynamic meshing feature used for the surge and heave oscillations were validated using experimental data of the NACA 0012 airfoil.

The effect of phase shift between the input wind gust signal and the model oscillating signal was analysed. The force and moment responses have shown phase shift compared with the 0 phase shift case. A maximum phase shift in the drag response has occurred for the 180-degree response. This was because of the maximum phase difference between the peak inputs for the 180-degree condition. The phase shift for lift and moment responses is almost linear, and the 0-degree response lags behind the other three responses. The drag response has demonstrated a non-linear change in its phase shifts. The effect of wind gust amplitude on the aerostat response was analysed for an AoA of 10 degrees. The phase shift in the responses with respect to the 10 % case was analysed. A linear variation was observed in the response amplitude with the gust amplitude. The phase difference among the responses has shown a reduction with the rise in the gust amplitude. The 30 and 50 % responses were lagging behind the 10 degree case for the surge, heave, and pitch oscillations. The effect of the AoA was analysed by considering four different angles, which were selected so as to include almost all the operating conditions of the aerostat. Drag response remained almost constant for the surge, heave, and pitch oscillations for the AoA considered. The effect of gust length was analysed by considering an AoA of 10 degrees and a 30 % wind gust amplitude. The amplitude of the responses has shown variation with the time period, whereas the phase was negligible compared with the previous set of simulations.

The proposed low-altitude tethered aerostat was intended to deploy in open atmospheric conditions with extreme wind conditions. The feasibility analysis of the aerostat was investigated by considering different wind conditions and motion conditions of the aerostat using CFD-based analysis. As the application demands an accurate position and altitude for the aerostat, the design and development of a control system are necessary. A mathematical model is the most significant stage for such a control system analysis. The current research considered the investigation of the mathematical model of the tethered aerostat under steady wind conditions.

A decoupled longitudinal model of the tethered aerostat was investigated in the current research. The lateral dynamics were assumed to be negated by the control surfaces. The model parameters of the aerostat (added mass terms and the stability derivatives) were estimated using a CFD-based methodology. The geometrical parameters of the aerostat were calculated from the CAD model. As the model parameters were obtained for a mean wind velocity of 7.5 m/s, the presented model was expected to work around the mean velocity.

A two dimensional model of the tether was investigated in the current research. The dynamics of the model were investigated by discretising the tether into a number of segments. The forces involved in the dynamics were distributed among these segments. The presented tether model was validated using experimental data available in the open literature. The tether length measured for different wind velocities was compared with the experimental data, and the results from current research were promisingly similar to those of the experimental data. The number of segments (or the length of each segment) for the tether model was selected after conducting an analysis of the trade-off between the accuracy of the model and the computational cost. The tether profile (two dimensional coordinates of the tether segments) was analysed for different wind amplitudes and different tether configurations, such as a fixed tether length and a variable tether length. The blow-by (horizontal drift in the aerostat position due to wind) of the aerostat and the variation that occurs in the aerostat altitude were investigated for a range of wind velocities.

The simulation of the tethered aerostat model was conducted using MATLAB 2020B. The simulation started with the initial model parameters of the aerostat and the tether. The tether equations were simulated first by considering a recursive approach so as to calculate the tether tension, drag, and position of each tether segment, starting from the top segment to the last segment. Once the tether profile is simulated, the total tether length, weight, tether tension, angle of the tether with the horizontal and position of the top segment were calculated and coupled with the aerostat equations of motion. ODE45 solver was used for solving the aerostat equations of motion. The response of the tethered aerostat was obtained for the steady mean wind velocity specified by NIWE. The response showed that the aerostat came to rest after 500 s after an initial overshoot in the forward and vertical velocities. The pitch angle response was negative, and it was due to the fact that the stability derivatives were estimated by considering the aerostat centre of volume as the pitching moment centre instead of the tether confluence point. Even though the presented model was supposed to be working around the specified mean wind velocity, an investigation into the effect of wind velocity was carried out. Tuning of some of the aerostat model parameters helped to achieve the response of the aerostat for different wind velocities as expected.

7.3 Thesis contributions

The major research contributions of this thesis can be summarised as follows:

- Wind gust analysis on the aerodynamic characteristics of an aerostat using discrete and continuous wind gust models using CFD for different test conditions.

- Scaling methodology for generalising the aerostat responses for various test conditions, including AoA, gust amplitude, time period, and aerostat shape.

- Stability derivatives estimation methodology for aerostats undergoing steady as well as unsteady wind conditions.

- A longitudinal mathematical model for a single-tethered aerostat-based on stability derivatives.

7.4 Scope for future work

The aerostat shapes considered in this thesis are taken from the literature. The effect of the aerodynamic shape at the nose portion of the aerostat is predominant in the investigation.

There is scope for shape optimisation for the aerostat based on the required lift-drag performance. The scaling methodology proposed can be made more general by investigating different curve fitting algorithms and geometric formulations.

The wind gusts considered in this thesis are horizontal gusts (head-on gusts). There is scope for investigating the effect of a vertical gust on the aerodynamic performance of the aerostat. The interaction and influence of the combination of horizontal and vertical wind gusts on the aerostat are also an open problem.

The stability derivatives of the aerostat were investigated for the longitudinal dynamics in this thesis. Lateral dynamic stability derivatives can be investigated by considering vertical gusts for the analysis. The possibility of developing a machine learning model/framework for predicting the aerodynamic behaviour of the aerostat can be considered an open problem. It will help reduce the number of CFD simulations and their time.

The mathematical model of the single-tethered aerostat was developed by assuming the leashes as rigid. A dynamic model for the leashes can be investigated, which will add multiple control DoF to the system. Investigations of different control algorithms to improve the operation of the tethered aerostat can be considered as well.